Dynamic Modelling for Supply Chain Management

Adolfo Crespo Márquez

Dynamic Modelling for Supply Chain Management

Dealing with Front-end, Back-end and Integration Issues

Springer

Adolfo Crespo Márquez, PhD
Department of Industrial Control
School of Engineering
University of Seville
Camino de los Descubrimientos, s/n
41092 Seville
Spain
adolfo.crespo@esi.us.es

ISBN 978-1-4471-5757-1 ISBN 978-1-84882-681-6 (eBook)
DOI 10.1007/978-1-84882-681-6
Springer London Dordrecht Heidelberg New York

British Library Cataloguing in Publication Data
A catalogue record for this book is available from the British Library

Cover design: eStudioCalamar, Figueres/Berlin

Printed on acid-free paper

Springer is part of Springer Science+Business Media (www.springer.com)

To the University of Seville

Foreword

The employment of supply chains is hardly a new concept. The ancient Egyptians, for example, developed relatively sophisticated supply chains in the construction of their pyramids. The Persian Empire, from 550 to 330 BC, was the largest empire of the ancient world and its success was due, to a large degree, to the design of its supply chains. The role of supply chains in the development of the Roman Empire was just as important, if not more so. Throughout recorded history, battles and even wars have been won or lost as a consequence of supply chain management.

Today the importance of supply chains and supply chain management is perhaps even more important. Future prospects for the growth and prosperity of firms and countries will largely depend on the design and oversight of their supply chains. In spite of this fact, the majority of supply chains in existence at this time have, like Topsy (a character in the novel Uncle Tom's Cabin) "just growed".

One of the more frustrating encounters in my career centered around a certain hi-tech's firm supply chain and business processes. The firm devoted substantial resources to the design of its products, the reduction of product defects, and the reduction of manufacturing expenditures. Little attention, however, was paid to the structure and oversight of its supply chain and the policies and procedures employed in its operation. These aspects of the firm's business simply evolved over time according to the whims and wishes of its management. As a consequence the dissatisfaction of the firm's customers grew and its market shared was diminished. Despite the production of truly outstanding products, the fortunes of the firm in question went into a rapid decline. Today the firm no longer exists.

During the later part of the twentieth century effort was devoted to the achievement of a better understanding and more scientific basis for supply chains. The growth of articles, books and courses on supply chains would appear to have grown at a near exponential rate. Unfortunately, too many of these efforts proposed concepts that relied more on principles, guidelines and slogans than on the provision of a comprehensive and scientifically sound approach.

Adolfo Crespo Márquez has written a book that is both practical as well as a tome based on science. His work replaces intuition, overly simplistic static supply chain models and sometimes questionable – or at least impractical –

principles with a holistic methodology that encompasses and explicitly considers the complexity and variability found in the real world. More to the point, the methodology employed satisfies a fundamental requirement of science; it is repeatable.

Adolfo Crespo Márquez presents a perspective of supply chains that incorporates the relatively recent "front-back" organisational model – a model that departs from the traditional product division perspective. The front end addresses those portions of the organisation and its business processes that deal with sales and marketing, organised according to customer type. The back end portion of the model encompasses the units that deal with research, development, and the methods and processes of manufacturing. These, in turn, are organised by product or technology type.

While the "front-back" model concept has been known for more than three decades, its employment has not received the reception it is due. This is, in great part, because such models are difficult to make work. Such a model requires that a firm must organise one way in the front end and yet another at the back end – and then successfully integrate both structures.

The book's author overcomes this obstacle to the adoption of the "front-back" model through the introduction of a systematic process for the employment of dynamic simulation models that may be used to both structure and analyse such models.

This book is a valuable addition to the literature and will be useful to both practitioners and analysts.

Dr. James P. Ignizio
Founder and Principal
The Resource Management Institute

Preface

This manuscript deals with specific problems, in different functional areas, related to the pursuit of organisations in becoming more customer-focused. These problems appear in many corporations migrating from product divisions to a "front-back" organisational model. Basically, this model designs the organisation considering two units called "front end" and a "back end", as follows:

- The "front end" units deal with units handling sales and marketing, and are organised according to customer type. These units are able to offer specific integrated solutions to customers.
- The "back end" units deal with research, development and elements of manufacturing. They are are organised by product or technology type, and they are able to provide the modular elements to be combined into solutions.

Front-back structures are notoriously difficult to make work. The problem is organising one way at the front, one way at the back, and somehow linking or integrating the two together. The models that will be presented in this book try to help in this process. They show how solutions to these problems can be found through the use of appropriate dynamic simulation models.

This work concentrates on hi-tech supply chains and networks problems inside a front-back organisational model. As the reader may guess, these problems are related to many different topics of management science like marketing, operations, financial and risk management, etc.

Special challenges are faced in trying to find an appropriate solution by using models and the reader will realise how the need for an interdisciplinary approach when using dynamic modelling is compelling. The work is divided into five major parts:

- Part I. An introduction to dynamic modelling for supply chains.
- Part II. Modelling front-end issues in SCM.
- Part III. Modelling back-end issues in SCM.
- Part IV. Modelling integration issues in SCM.
- Part V. Dynamic Modelling Projects.

Each of these five parts covers different contents with the following intentions:

- Part I of the book is an introduction to the modelling methodology. Main concepts and procedures to build dynamics models properly will be reviewed. Literature concerning works dealing with dynamic modelling and supply chain management topics will also be reviewed.
- Part II is a review and proposal of dynamic modelling options to connect customer value to business targets. This is carried out by explaining how to characterise the target market by formalising what are often informal but deeply held beliefs concerning what drives their customers' purchase decisions.
- Part III discusses and explains experiences in modelling different types of supplier contracts to accomplish varying degrees of security and flexibility. Attention is focused on business dynamics based on current best practices in portfolio management, as shown by leaders in volatile high-technology businesses. This part of the book also deals with manufacturing issues and problems that can be explored by using this methodology.
- Part IV reviews and discusses the operational and financial effectiveness of existing virtual tools used in supply chain integration. It illustrates how dynamic modelling may help to obtain a comprehensive model of supply chain integration, a modelling effort that can be used for the analysis of the effectiveness of various levels of integration, as well as for the assessment of the importance of the sequence in which virtual collaboration tools are adopted in supply chain integration. This part of the book also deals with cultural diversity issues and problems that can be explored by using dynamic modelling.
- Part V of the book includes various experiences and captured learning, that can be useful in the process of presenting, opening, developing or closing dynamic modelling projects.

Most of the models in this book are presented formally and the reader may easily implement them regardless of the software she/he may want to use. Models cover many different topics, all related to organisational change and improvement.

All the models are preceded by one or various case studies. A case study introduces the reader into the topic and problem, then tries to reveal and show, somehow, the business "call for action".

Escuela Superior de Ingenieros
Isla de la Cartuja, Sevilla, Spain
December 2009
Adolfo Crespo Márquez

Acknowledgements

I wish to thank specific people and institutions for providing their help, making the publication of this book possible.

The University of Seville granted me a visiting scholarship to Northwestern University (Evanston Il.) in 1996. During that scholarship I met most of the people and institutions that have made this book possible. The University of Seville also allowed me to travel in many ocassions to the USA during the years 1996–2003, to complete several modelling efforts and to follow and offer different workshops and seminars. Materials and knowledge gained during that time now serve as part of subjects that I am currently teaching: "Continuous Simulation" in the School of Engineering of Seville; "Modelling Manufacturing Systems" in the School of Engineering of the Swansea University; and "Innovation Marketing" in the Vienna University of Technology . I dedicate this book to the University of Seville in gratitude for all these wonderful opportunities for personal and professional development.

For many years Rafael Ruiz Usano has been the Head of the Research Group "Organización Industrial" at the School of Engineering of the University of Seville. Within this group, several colleagues have found an amicable and friendly working atmosphere where the area of dynamic modelling could develop. I thank Rafael for his support.

Deb Campbell and Greg Jacobus (both from Hewlett-Packard, in Palo Alto during the summer of 1996) offered me, while I had a visiting scholarship in Northwestern University, the opportunity to join some dynamic modelling efforts at HP in the late 1990s. I especially have to thank Deb for many things I could learn about the hi-tech corporations, the complex model-building processes, the overall model process facilitation, or the opening and closure of modelling projects. Deb also co-authored several papers for the International System Dynamic Conference. All of those things were very important for this book, as well as an excellent personal relationship with Deb and her family during those years.

Carol Blanchar (from Conexo, Santa Clara, CA. USA) provided help with several case studies related to her consulting activities with organisations in different parts of the USA. Her support was especially valuable with material regarding contract portfolio analysis and customer value analysis among other topics. Carol also co-authored several papers related to front-back models topics in IJPE and DSS Journal as well as an international patent related to a methodology to evaluate new investments in hi-tech products. I consider Carol a very knowledgable person and, together with her husband David, even better friends.

Andre Kuper provided extremely valuable input regarding tools to improve velocity and flexibility in supply chains. In the late 1990s André Kuper aligned people with new business models for Hewlett-Packard. Prior to his work with HP, Andre worked at the Applied Superconductivity Lab at University at Twente and at Accenture ECC in Enschede, The Netherlands.

Jim Ignizio, besides writing the foreword of this book, was, in 2004, the Director of the MOSAIC3 project in Intel Corporation Fab 11X ME (Albuquerque, New Mexico, USA). Jim invited me and gave me the opportunity to learn about the Fab and to apply dynamic simulation techniques to certain specific manufacturing problems. Jim also co-authored a paper in PPC Journal related to dynamic simulation models to improve maintenance scheduling in semiconductor fabs, which serves as basis for Chapter 10 of this book.

Venu Nagali is a distinguished technologist and HP Procurement Risk Management (PRM) leader. His presentation of this approach to the Supply Chain Management Council serves as a basis, together with materials provided by Greg Jacobus, of an introductory case to the dynamic contract portfolio management models.

Sharone Zehavi was in 2003 President and CEO of Global Factory Inc. He introduced me to several compelling applications and case studies allowing supply chain partners to communicate in a common language through cross-corporate application integration. Some of those ideas are included in the case study presented in Chapter 11.

Kevin McCormack (from DRK Research) provided permission to use some of the figures related to PRM in Chapter 8.

Salvatore Cannella and Elena Ciancimino, post-graduate students from the University of Palermo, now at the the University of Seville, provided a very valuable literature review and simulation efforts in Chapter 14 dedicated to constrained SCs.

As well as people contributing different material and valuable knowledge to this work, there are also other colleagues who reviewed many of the concepts and case studies in the book. In this sense, I would like to thank Prof. Jatinder (Jeet) Gupta from the University of Alabama Huntsville and Prof. Carmine Bianchi from the University of Palermo in Italy.

The funding from the Spanish Ministery of Science and Education during the time this book was written (Research Projects DPI:2004-01843 and DPI:2008-01012) made many things related to this work possible.

Finally, a special thank you to the author's wonderful family: Lourdes, Lourdes Jr, Adolfo Jr and Gonzalo, who offered him their love, support and precious time, enabling this work to be accomplished.

To all of them, thanks.

Contents

Part I An Introduction to Dynamic Modelling for Supply Chains

Part IV Modelling Integration Issues in SCM

Part I

An Introduction to Dynamic Modelling for Supply Chains

1

On the Definition of Dynamic Simulation

1.1 An Introduction to Dynamic Simulation

1.1.1 Systems, Models and Simulation

The Webster Dictionary defines a *system* as a group of independent but interrelated elements comprising (acting as) a unified whole; it can also be defined as a process for obtaining an objective.

A *model* is defined as a representation of something, a simplified description of a complex entity or process. Therefore we can generate models of systems.

Modelling refers then to the process of generating a model as an abstract representation of some real world entity, process or system.

Typically a model will contain only the significant features or aspects of the item/system in question, and two models of the same item/system may differ quite significantly. This may be due to differing problems to be solved by the model's end user (one user may be interested in aspects of the item which are quite separate from those of another user). For this reason it is critically important for any end user to understand the problem to solve, the original purpose, or the application for the model.

In this book we deal with mathematical models; these are abstract models, mathematical structures, using mathematical language to describe the behaviour of a system. A mathematical model usually describes a system by a set of variables and a set of equations that establish relationships between the variables. The values of the variables can be practically anything; real or integer numbers, Boolean values, strings, etc. The variables represent certain properties of the system, for example, measured system outputs often in the form of signals, timing data, counters, event occurrence (yes/no). The actual model is the set of functions that describe the relations between the different variables.

We can find mathematical models falling, for instance, within some of the following categories (taken from Webster, Britannica & Sci-Tech dictionaries):

- *Linear (vs nonlinear).* Mathematical models are usually composed by variables and operators which can be algebraic operators, functions, etc. If all the operators in a mathematical model present linearity, the resulting mathematical model is defined as linear. A model is considered to be nonlinear otherwise.
- *Deterministic (vs probabilistic or stochastic).* A deterministic model is one in which every set of variable states is uniquely determined by parameters in the model and by sets of previous states of these variables. Therefore, deterministic models perform the same way for a given set of initial conditions. Conversely, in a stochastic model, randomness is present, and variable states are not described by unique values, but rather by probability distributions.
- *Dynamic (vs Static).* A dynamic model accounts for the element of time, while a static model does not. A dynamic mathematical model is a model that describes how a system changes in time and may have a variety of representations, from the traditional notations of mathematics to diagrammatic (we will use several representations of dynamic mathematical models in this book).
- *Others.*

Once we have a model representing a given real world system, S*imulation* is attempting to predict aspects of the dynamic behaviour of the system the model represents (see the Free On-line Dictionary of Computing at [13]).

Traditionally, the formal modelling of systems to predict their behaviour has been via a mathematical model which attempts to find analytical solutions enabling the prediction from a set of parameters and initial conditions. For many systems, however, simple closed form analytic solutions are not possible. This is the point at which *computer simulation models* come into play. *Computer simulation* is often used as an adjunct to, or substitution for, modelling systems for which these analytic solutions are not possible. It generates a sample of representative scenarios for a model in which a complete enumeration of all possible states would be prohibitive or impossible.

In this book we will see how *computer simulation modelling* is extremely well suited to study systems that are dynamic and interactive as well as complicated. This technique has been in use in management science since the early 1950s and its methods have gradually evolved alongside general developments in computing science ever since [1].

An important aspect to take into account is that "simulation should imitate the internal processes and not merely the results of the thing being simulated". That is to say that a simulation model should somehow capture the *structure of a system* in order to predict aspects of its *behaviour*, with the purpose of *solving a certain problem.*

1.2 Time Handling in Dynamic Simulation Models

1.2.1 Type of Dynamic Computer Simulation Models

Dynamic mathematical models used in computer simulation are typically represented with differential equations (the relationship involving the rates of change of continuously changing quantities modelled by functions) or difference equations (relating a term in a sequence to one or more of its predecessors in the sequence). There is a clear reason for this that is related to the nature of the system being modelled. Some industrial systems or processes, like many process plant processes, occur continuously in time. Others, such as certain manufacturing processes, occur more discretely in time. Even though data collected from continuous processes are by necessity taken at discrete time intervals, model predictions based on these data assume temporal continuity and are commonly written in the form of differential equations. By contrast, discrete-time processes are modelled using difference equations, equations that take into account the discontinuous nature of these processes.

1.2.2 Difference Equations in Discrete Time Models

Difference equations are used in systems where change occurs at discrete points in time. Difference equations suppose that future values of variables of a system are a function of the current and possibly past values.

For instance, a first-order difference equation, given below, supposes that the next period value is only a function of the current period value:

$$x_{t+1} = f(x_t) \tag{1.1}$$

where $f(x_t)$ may be either a linear or nonlinear function, and the starting value x_0 is needed for the equation to be solved. A general k-order difference equation takes the form

$$x_{t+k} = f(t, x_t, x_{t+1}, \ldots, x_{t+k-1}) \tag{1.2}$$

Obviously, for a k-order equation we need k-1 starting values – $x_0, x_1, \ldots, x_{k-1}$ – to determine x_k. Again, $f(t, x_t, x_{t+1}, \ldots, x_{t+k-1})$ may be either a linear or nonlinear function.

1.2.3 Differential Equations in Continuous Time Models

Another way to model dynamics is to assume that change occurs continuously rather than at discrete points in time. The continuous time analogue to difference equations are differential equations that can be written as

$$\frac{dx}{dt} = f(x,t) \tag{1.3}$$

where $f(x,t)$ can be a linear or nonlinear function.

As with difference equations, a system of differential equations can be specified to represent the behaviour of several and interacting variables over time. Various methods can be used to generate solutions to systems modelled by differential equations. Derivatives are the mathematical formalism for describing continuous change. The differential equation which embodies a model provides the values of these derivatives at any particular time point; calculus or a computer can then be used to move the state of the model forwards in time.

Continuous models have the advantage over discrete time models in that they are more amenable to algebraic manipulation, although they are slightly harder to implement on a computer.

1.2.4 Computer Simulation Models Time Advance Methods

The actual process of computing the model state and producing the state values as the simulation time is advanced in the computer is called model execution [2].

A key design element in model execution is the time advance mechanism [3]. Most common time advance mechanisms are:

- Time-stepped. Time is advanced in fixed time increments and the system state is updated (recalculated) at each increment.
- Discrete-event. Different part of system state evolve at their own timescales, using the concept of events. Each event signals the specific instant in simulation time at which a particular part of the system is to be updated.
- Time parallel. In this case simulation time is partitioned in multiple segments, and each segment is executed independently from each other.

1.2.5 Executable Timelines

The model execution normally requires the consideration of three different time axes [2]:

- Physical time. Time in the physical system that is being modelled. For instance, and assuming units of time in weeks, from week 1 to week 45 of the year 2008.
- Simulation time. Representation of the physical time for the purpose of the simulation. Corresponds to the simulated time period of the physical system. For instance, number of weeks since the beginning of the year 2008.
- Wallclock time. Ellapsed real time during execution of the simulation, as measured by a hardware clock. For instance, number of miliseconds of computer time during execution.

1.3 Deterministic and Stochastic Simulation

Many of the models included in this book contain stochastic elements. The first implication of this is the need for a more careful treatment of model results [4]. The result of every model simulation (experiment) has to be considered as samples and these samples depend on the random number streams used to produce them. Different random numbers will transform into different samples, and simulations will produce different results. In order to reach confidence in these results it is important to produce a certain set of results (samples) and that those results are analysed using suitable methods. The greater the number of results (runs) the greater the confidence that the results are representative. Another important feature when using stochastic simulation is the fact that when comparing different policy options each option should be compared using the same random numbers. This ensures fair comparison of alternatives. A third important point [1] is that design of experiments is required. Analysis of experiments is a statistical field which may require modeller attention.

1.4 Dynamic Modelling Methodology and Tools

1.4.1 System Dynamics

System Dynamics is ([12], the official web page of the System Dynamics Society) a methodology for studying and managing complex feedback systems, such as one finds in business and other social systems. In fact it has been used to address practically every sort of feedback system. While the word system has been applied to all sorts of situations, feedback is the differentiating descriptor here. Feedback refers to the situation of X affecting Y and Y in turn affecting X perhaps through a chain of causes and effects. One cannot study the link between X and Y and, independently, the link between Y and X and predict how the system will behave. Only the study of the whole system as a feedback system will lead to correct results.

The basis of the method is the recognition that the structure of any system — the many circular, interlocking, sometimes time-delayed relationships among its components — is often just as important in determining its behaviour as the individual components themselves. There are often properties-of-the-whole which cannot be found among the properties-of-the-elements; in some cases the behaviour of the whole cannot be explained in terms of the behaviour of the parts.

The methodology:

1. identifies a problem;
2. develops a dynamic hypothesis explaining the cause of the problem;
3. builds a computer simulation model of the system at the root of the problem;

4. tests the model to be certain that it reproduces the behaviour seen in the real world;
5. devises and tests in the model alternative policies that alleviate the problem; and
6. implements this solution.

1.4.2 System Dynamics Modelling Tools

In order to develop steps 2 and 3 of the System Dynamics methodology, we can find some very practical tools such as the casual loop diagrams and the stock and flow diagrams:

- *A causal loop diagram (CLD)* is a diagram that aids in visualising how interrelated variables affect one another (see Figure 1.1). The simple diagram notation of nodes and lines identifies the important variables in a system and how they interact. The CLD presents an easily understood conceptual model of how the system works, but even more important is the fact that CLD provides a language to communicate, to interact and to exchange points of view about the structure of the system we are about to model.

  The diagram itself consists of a set of nodes representing the variables connected together. The relationships between these variables, represented by arrows, can be labelled as positive or negative (which can be denoted with a "+" or "-", respectively). Positive causal links means that the two nodes move in the same direction, *i.e.* if the node in which the link start decreases, the other node also decreases. Similarly, if the node in which the link starts increases, the other node increases. Negative causal links are links in which the nodes changes in opposite directions (an increase causes a decrease in another node, or a decrease causes an increase in another node).

  The causal effect between nodes determine positive reinforcing loops or balancing loops (which can be denoted with an "R" and "B", respectively). Reinforcing loops (which can be denoted with an "R") have an even number of negative links (zero in the simple example above) and balancing loops an uneven number.

  Identifying reinforcing and balancing loops is an important step in System Dynamics because it helps to identify *reference behaviour patterns*, *i.e.* possible dynamic behaviours of the system. The first article on System Dynamics, written by Jay W. Forrester, appeared in *Harvard Business Review* in 1959 [5] and used principles of information-feedback control to explain how aggressive advertising by a company could create workload fluctuation on the shop floor. This approach to modelling management processes introduced the notion that the dynamics of an industrial system arises as a result of its underlying structure. The basic structural element is the feedback loop; the underlying structure refers to the collection of interacting feedback loops comprising the system. This linkage between structure and behaviour remains the guiding principle for practitioners of systems dynamics. These practitioners associate a reinforcing loop with an exponential increase/decrease while balancing loops are associated with reaching a plateau. System delays (often denoted by drawing a short line across the causal link) may cause the system to fluctuate. In this way, behaviour of the systems can be explained through the analysis of feedback loops, their *gains* and *delays*, over the simulation time.

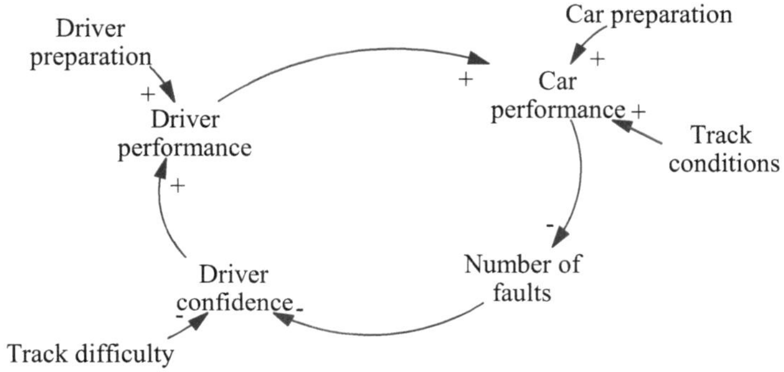

Figure 1.1. Sample causal loop diagram (CLD)

- Stock and flow diagrams (SFD) — or level and rate diagrams (LRD) — are ways of representing the structure of a system with more detailed information than is shown in a causal loop diagram. Stocks (levels) are fundamental to generating behaviour in a system; flows (rates) cause stocks to change. Stock and flow diagrams contain specific symbols and components representing the structure of a system. Stocks are things that can accumulate — such as inventory — and are represented with boxes.

  Flows represent rates of change and they are expressed by decision functions — such as reductions in inventory through sales — and they are represented or drawn as valves. These diagrams also contain "clouds", which represent the boundaries of the problem or system in question; auxiliary variables, etc. Systems are composed of interconnected networks of stocks and flows, including many information channels, which connect the levels to the decision functions. Modellers must be able to represent the stock and flow networks of people, material, goods, money, energy, etc. from which systems are built. Stock and flow diagrams are the most common first step in writing the executable code of a simulation System Dynamics model because they help to define types of the variables that are important in causing behaviour. Therefore we can say that stock and flow diagrams provide a bridge from conceptual modelling to assigning equations to the relationships between variables.

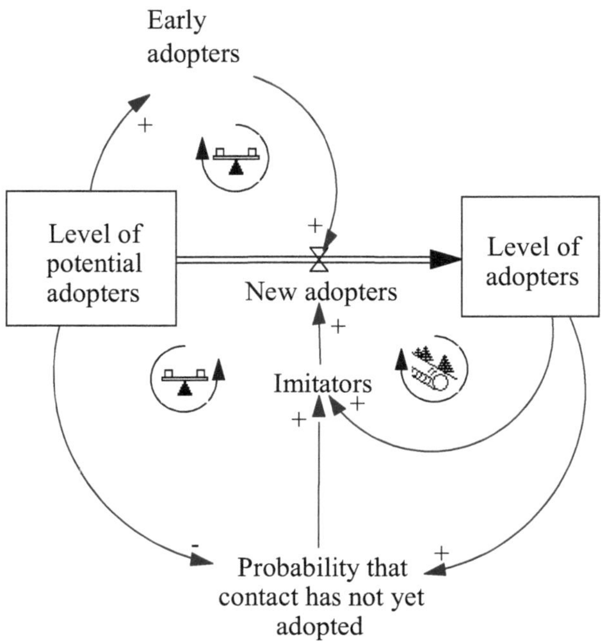

Figure 1.2. Sample Stock and Flow Diagram (SFD)

Figure 1.2 depicts a very simple structure of a reservoir or level, with an inflow and an outflow. To specify the dynamic behaviour, a system of equations is defined. It consists of two types of equations, which correspond to levels and decision functions (rates). Equations control the changing interactions of a set of variables, as time advances. The continuous advance of time is broken into small intervals of equal length *dt*. For example the equations describing the state of the levels in Figure 1.2 is

$$Level(t) = Level(t - dt) + dt \cdot \left[Decision\ \ Flow\ \ 1(t) - Decision\ \ Flow\ \ 2(t)\right] \quad (1.4)$$

$$Level(t_o) = Lo \quad (1.5)$$

Levels in Figure 1.2 at time *t* depend on its value at time *t-dt* and the value going in from decision function 1 minus the value going out to decision function 2. Notice that it is necessary to give the initial value of it to solve this equation.

There will be as many equations as variables. To determine the variables' behaviour, the differential equations system is integrated. This can be done with software that supports this and which uses different numerical integration methods.

Sometimes, however, it has been found that SFD is a very specific tool "only for analyst" and/or "model builders". It may create confusion when used as a general purpose tool for model building with business teams, etc. There exists some empirical research [6] showing that even highly educated people may have difficulties in discerning between stocks and flows.

1.4.3 System Dynamics Software Tools

An enormous value of modern system dynamic modelling tools is that they facilitate the process of capturing models of the underlying behaviour structure of organisational systems. The modelling software available on the market today greatly contributes toward achieving that objective by allowing model builders to concentrate on conceptualising the system rather than on the technicalities of model building [7].

The most popular commercial software packages are Powersim [8], iThink [9] and Vensim [10]. All three provide the following basic capabilities:

- Drawing the model (CLD and/or SFD) using an interface. Modelling elements from the toolbar are dragged and dropped onto the white area to create the structure. For stocks, initial values need to be specified. Decision rules for the flow variables and converters are written by entering the dialogue box.
- Building the model code to be executed in the computer. Decision rules for the variables are written by entering dialogue boxes, which incorporate a rich set of built-in functions allowing mathematical representation of most real-life situations.
- Simulating the model with different values of certain model parameters.
- Publishing the results both as table and graph.
- Performing sensitivity analysis and publishing comparison of run results.

Beyond these basics, each package also provides additional features that are now laid out and that may make each one suitable for particular modelling situations:

- *Vensim*® (Vensim is a registered trademark of Ventana Systems Inc.) provides high rigour for writing model equations. It adds features for tracing feedback loops. In addition "Causes Tree" and "Uses Tree" features help in debugging the model. Vensim also provides very powerful tools for multiparametric simulation results optimisation which allows the analyst to validate results and model structure as well as to determine most convenient policy options by parametrising these policies.
- *iThink*® (iThink is a registered trademark of Isee Systems Inc.) provides a multi-level modelling interface that allows for separating out the user interface, the stock and flow model and the equations into three different levels. The interface level can be used to show an overview of the

system, the causal loop diagram and model outputs. The model tracing facility provides an easy way to navigate through the feedback loops and learn about the reasons behind the dynamics. iThink in recent times has been used to build multimedia games with the aim of providing managers an experimental set up for experiential learning [7].

- *Powersim* (Copyright Powersim Software AS ©) comes with the powerful feature of adding user written functions. This can become useful in modelling situations where new concepts (e.g. fuzzy logic) need to be incorporated. Latest versions of Powersim can build reusable model components that can be plugged in without much difficulty [7].

1.5 Model Validation *vs* Usefulness

In a practical sense, analysts are concerned with usefulness rather than validity of the models. Does the model serve the purpose for which it was intended? Is it helpful? Therefore, the developer's or user's purpose must be kept in mind in evaluating a model's usefulness, or validity.

The selection of an appropriate level of detail, problem boundaries, and similar considerations constitute the "art" aspect of dynamic simulation model development. Many times, validity or usefulness lies in the subjective view of the user. We think of models as valid when they can be used with confidence. With this in mind, how can one gain confidence in dynamic simulation models? Here we lay out some interesting aspects to be considered [11]:

- Because the foundation for model behaviour is the model's structure, the first test in validating a model is whether the structure of the model matches the structure of the system being modelled. Structure exploits judgment, experience, and intuition. Data plays a secondary role.
- The model's parameter values are a specific area for testing. Parameter values in a model often may be tested in a straightforward manner, e.g., against historical data. However, in dynamic simulation models of socioeconomic systems the desired data may be unavailable, in an inappropriate form, or incorrect. There may be elements that are not usually quantified, but that are critical to the system being modelled. These elements must be included in the model. The point is that dynamic simulation model parameter values, from whatever source they may be derived, are subject to a rigorous and demanding environment. These values contribute significantly to confidence in the model when the specified parameter values are reasonable and consistent with whatever supporting data might exist.
- Model boundaries must match the purpose for which the model is designed, if the model is to be used with confidence: that is, the model must include all of the important factors affecting the behaviour of interest. In practice, boundaries tend to shift as the developers' and users' understanding of a problem evolves with the model's development. As

model purpose shifts, changes in the model's boundaries may be required.
- A less obvious test relating to model structure involves the effects of extreme conditions. The ability of a model to function properly under extreme conditions contributes to its utility as a policy evaluation tool as well as user confidence.
- Together with the dynamic, rather than the static, nature of the simulation, these characteristics have shifted emphasis from more traditional, statistical tests to the kinds of tests described in the previous points – whole model tests that engage all the model variables and their relationships in the testing process.

1.6 Dynamic Modelling Approach Followed in this Book

According to the concepts explained in the previous sections of this chapter, dynamic simulation models that will be presented in this work can be characterised as follows:

1. They will be nonlinear models, i.e. their variables and operators will not present, as a general rule, linearity.
2. Some of the models will be stochastic. Randomness will be present and variable states will not be described by unique values.
3. Difference equations will be used to formalise the models, i.e. future values of variables will be expressed as a function of the current and possibly past values.
4. The time advance method will be time-stepped, i.e. time will be advanced in fixed time increments and the system state will be recalculated at each increment.
5. The time that the physical system is modelled – physical time of the simulation – will depend on the purpose of the specific analysis to carry out. For instance, in Chapter 9, the portfolio of supplier contracts analysis assumes units of time in weeks and the analysis is done for 104 weeks. In Chapter 10, time units selected to simulate the wafers toolsets are minutes. In that case a total of 40,000 min are considered appropriate for the simulation to show the impact of different maintenance scheduling policies.
6. An additional consideration here is that, for some examples and cases presented in this book, the systems that are simulated change their state at fixed physical time intervals. For instance, most of the real supply chain management systems modelled in this book, related to the hi-tech industry, considered a weekly update of processes such as ordering, shipment, invoicing, etc. That means that current management systems running the business consider a weekly update of the information, as a review period, in their decision-making processes. For these particular scenarios, the simulation time clock can be advanced at fixed time

intervals (1 week), and the state of the simulation model is updated at the same recurring regular intervals as the physical system.

7. Simulation will be, for the general case, stochastic and the results of the different experiments will be considered as samples. Most of the time confidence in these results will be reached using suitable inference methods.
8. The modelling methodology to follow will be the one presented for System Dynamics in Section 1.4.1. In this case special attention is paid to the use of the different models as decision support systems. The use of system dynamic tools such as CLDs and SFDs will be at the discretion of the author and for each specific model. That means that these tools are not always used in the model building process. The reader will see that in some case studies more attention is paid to CLDs, or to SFDs or simply to the mathematical model formulation.
9. The simulation software tool used to build the models in this book is Vensim. This work benefits, on several occasions, from an interesting advantage of Vensim, that of the incorporation of a powerful optimiser based on a modified Powell method algorithm. This feature produces very fast convergence of the direct search technique when optimising solutions and without the requirement of gradient assessment in the different iterations. Having said this, it is important to remember that the mathematical formulation of the models in the book does not take into consideration the software used, *i.e.* Vensim code is not included in the models and the reader can build them regardless of the software tool used.
10. Regarding model validation, models built to deal with all the case studies presented in this book followed serious reality checks and validation procedures in the different organisations when they were being developed and later when they were being used. Some of them, as mentioned in Acknowledgements, became international patents after a broad and fruitful implementation in different companies and business units. Nevertheless, and as mentioned above, model validity lies in the subjective view of the user. In this sense, and as a general rule, impressions captured about the value provided by the models were always more positive during the modelling process than once the model was finished. Orienting modelling projects and case studies to foster organisational learning was always good practice. Understanding model structure and linking that to the model and therefore to system behaviour was found to be the key to that learning. Following this path, different modelling teams could achieve great results and some of these dynamic modelling projects were scored among the best valued projects in important corporations over several years.

1.7 References

[1] Pidd M, (2003) Tools for thinking. Modelling in management science. Chirchester: Wiley.
[2] Perumalla KS (2007). Model Execution. In: Handbook of Dynamic System Modelling. Edited by Fishwick PA. Boca Ratón: Chapman and Hall/CRC.
[3] Zeigler BP, Praehofer H, Kim TG, (2000) Theory of Modelling and Simulation, 2nd. Edition. New York: Academic Press.
[4] Law AM, kelton WD, (2001) Simulation Modelling and Anlaysis. 3rd. Edition. New York: McGraw-Hill international Editions.
[5] Forrester JW, (1959) Advertising: A problem in Industrial Dynamics. Harvard Business Review, 37(2).
[6] Booth-Sweeney L, Sterman JD, (2001) Bathtub dynamics: Initial results of a systems thinking inventory. System Dynamics Review, 16(4): 249–286.
[7] Dutta A, Roy R, (2002) System Dynamics. OR/MS Today. June. The Institute for Operations Research and the Management Sciences.
[8] Powersim. Powersim Corp, Bergen, Norway, http://www.powersim.com
[9] iThink. High Performance Systems Inc., Hanover, NH 03755, http://www.hps-inc.com
[10] Vensim. Ventana Systems Inc., Harvard, MA 01451, http://www.vensim.com.
[11] Shreckengost RC, (1985) Dynamic Simulation Models: How Valid Are They?. In: Self-Report Methods of Estimating Drug Use: Meeting Current Challenges to Validity. Division of Epidemiology and Statistical Analysis. National Institute on Drug Abuse. N I DA Research Monograph 57. Washington: U.S. Government Printing Office.
[12] http://www.systemdynamics.org/
[13] http://www.foldoc.org/

2

Current Supply Chains Management Issues

2.1 Current Issues in SCM

The concept "supply chain management" (SCM) is used in this book to refer to the means by which firms engage in creating, distributing and selling products [1]. That includes all cooperative efforts among members of the Supply Chain (SC) in order to reach higher market intelligence through a more precise market information gathering, product research, product development and design, and value analysis of the total system [2–4].

The term supply network will also appear in this text given the nature of the supply relationships at present, that is, non-linear flows, network-like systems and webs of suppliers and customers. Supply networks, as we will see, may become an extremely powerful competitive advantage for industrial organisations;

Notice that the presence of these supply networks becomes almost compulsory in cases where businesses have to deliver more value in new ways; to be faster to market, to become more flexible in responding to demand changes and to lower costs. In order to provide these higher service levels many companies have turned to external suppliers to provide them with capabilities that they themselves could no longer provide. Clearly, in such cases, real competition is no longer company *vs* company but SC *vs* SC.

With this in mind, what are the supply network capabilities needed for success in the marketplace? How do we integrate capabilities through contracts portfolios, unique products and/or services or relationships? These are the key strategic SC issues that will be addressed in this book.

2.2 SCM Issues and Related Problems

A vast list of SCM issues and related issues and problems can be found in the literature. Chandra and Grabis [5] summarise these issues and problems as shown in Table 2.1. They state that from this table it can be gleaned that SCM issues pose complex problems and that the SCM problem domain can be analysed at various levels of decomposition. On the first level, the overall problems of SCM consist of multiple sub-problems such as product design, network design, logistics management, customer service and others. Chandra and Grabis also define these problems as general and specific. Specific problems occur at the vertical direction of problem decomposition and deal with one particular issue, for instance, inventory management. General problems cross multiple specific problems horizontally. Dealing with problems requires solving multiple specific problems, for instance, ensuring customer service involves solving problems from logistics and sales areas.

Table 2.1. SCM issues, related problems and suggested problem-solving aproaches [5]

Supply chain issue and related problem	**Problem-solving Approach**
Distribution network configuration	Network flow optimisation
Inventory control	Forecasting and inventory management
Supply contracts	Global optimisation
Distribution strategies	Warehousing and transportation cost management
Supply chain integration and strategic partnering	Collaborative Planning, Forecasting and Replenishment (CPFR)
Outsourcing and procurement strategies	Managing risk, payoff tradeoffs with outsourcing *vs* buying
Information technology and decision support systems (DSS)	ERP implementation and Decision Support Systems (DSS)
Customer value	Statistical Process Control (SPC), Total Quality Management (TQM) and service level maximisation

It is clear that SCM involves dealing with multiple managerial and technical problems [4,6] highlighting several common issues that must be addressed for a supply chain to function effectively and efficiently.

2.3 Network Configuration and Competition

In a very interesting work Rice and Hoppe [7] studied, using the Delphi method, how supply chains may compete against each other. They considered and analysed three scenarios, since no single scenario provides a universally valid Characterisation of competition:

1. Competing as SC *vs* SC literally. Competition among groups of companies across the supply network competing as one entity, formally or informally. This competition applies when the following conditions

are present (notice that these conditions may apply only to one of the competitors):

- SC is a vertically integrated company, either competing against another similar vertically integrated company or against supply networks comprised of many companies;
- when the supply network is a highly integrated company with no common suppliers;
- when the supply network is comprised of companies that have sole-source relationships;
- when the industry is fragmented in such a way that there are no common strategic suppliers represented in more that one supply network, and most strategic suppliers are dedicated to one supply network.

2. Competing on supply network capabilities. Competition between individual companies competing on their internal supply network capabilities. Mainly competing on the effectiveness, efficiency and responsiveness of the network and on the network design used (for instance, applying innovative postponement production strategies, introducing new distribution channels, etc.). Network capabilities can be added or integrated (not copied).
3. Competing on supply network capabilities lead by a Channel Master. Competition centred on the single, most powerful company of a supply network (referred many times as the channel master). This scenario is commonplace in today's marketplace.

These three scenarios are considered not mutually exclusive; Rice and Hoppe presented cases of vertically integrated companies (ZARA) competing against Channel masters (The Limited) and against other parts of interconnected supply networks competing based on their network capabilities (The GAP).

In Figure 2.1 completely disconnected supply networks compete against each other with no overlaps at any tier (for example, automobile manufacturing supply chains of the US, Germany and Japan in the 1970s).

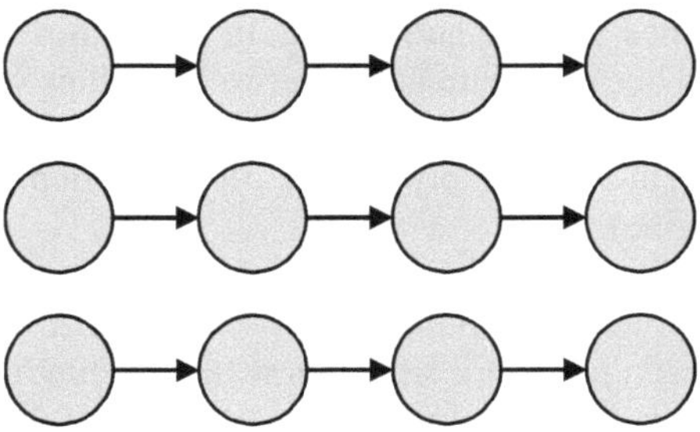

Figure 2.1. Completely disconnected supply networks (adapted from [7])

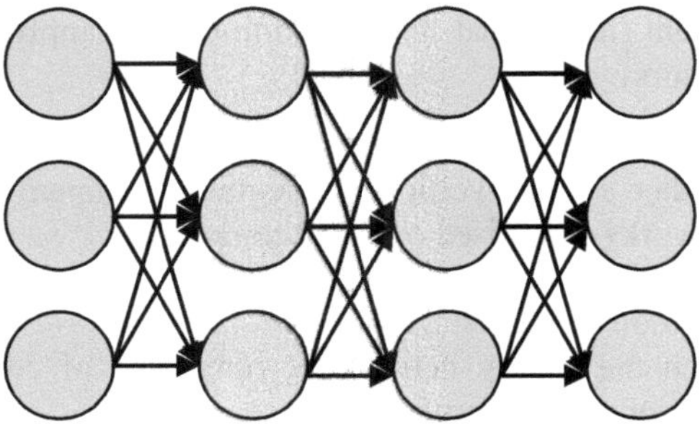

Figure 2.2. Completely overlapped supply networks (adapted from [7])

In Figure 2.2 each of the three networks overlaps with each other. Each company at every tier sells good to every tier (n+1) company.

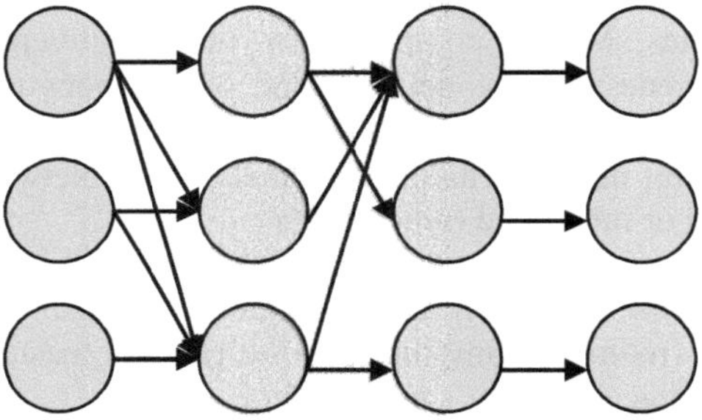

Figure 2.3. Partially overlapped supply networks (adapted from [7])

Competition in the hi-tech industry is, as in many other industries, somewhere between these two extremes (Figure 2.3) with some overlaps and some completely disconnected tiers within the networks.

Overlaps are common for commodity products procured efficiently from multiple members in the open market. For instance Dell and HP (and Compaq before the merge with HP) compete in modular product architecture and they have a fragmented supplier base creating significant overlap.

Also, as mentioned by Rice and Hoppe, in most cases many of the potential links are eliminated because there are closer relationships with some companies, depending on the nature of the product, price and capacity of the supply network (for instance, two hi-tech supply chains may overlap limited to memory, software and/or engine).

But once capabilities to be improved or developed are determined, we have to plan actions and tactics to put them to work. At this moment we have to understand that creating network strengths while meeting customer needs is not an easy job. This in fact requires higher and deeper levels of coordination among the companies in order to ensure that they create unique value. In the following paragraphs issues related to this phase will be reviewed.

2.4 Sharing Information Through ICTs

An important aspect to improve coordination among network companies is the evolution of the role played by information and communication technologies (ICTs). Researchers agree that sharing critical information, in context, in time, has been shown to reduce inventory dramatically and improves the performance at all SC levels (see Figure 2.4).

Shore [8] separates this evolution of ICTs in supply chain management into four stages:

- in the first stage, inter-organisational information exchanges travelled though the postal system or fax;
- using EDI, the second stage focused on the automation of information flows and the elimination of many labour intensive data entry and re-entry processes between retailers and suppliers;
- the third stage emphasises a more integrative strategy by implementing ERP systems;
- in the fourth stage, a supply chain is characterised by strategic supplier alliances with extensive two-way information flows.

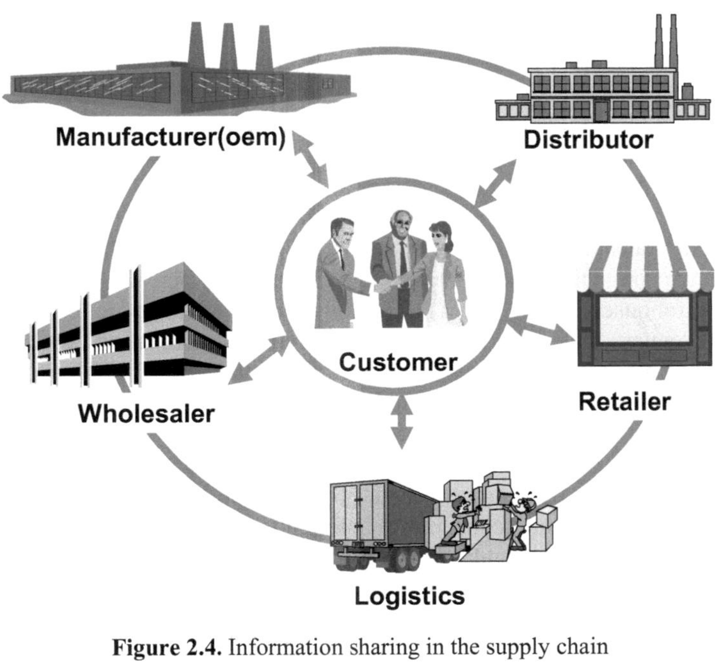

Figure 2.4. Information sharing in the supply chain

Collaboration by sharing information has joined the ranks of integration and automation as a hallmark of competitive advantage in the supply chains. C-commerce has been described [9] as achieving "...dynamic collaboration among employees, business partners and customers throughout a trading community or market...." The ability for businesses to "morph" into whatever the market needs them to be, in time, all the time, clearly means more than buy–sell transactions and auction events.

The benefits of c-commerce are similar to those achieved in the 1980s through concurrent engineering – reduced time to market, increased market share, and faster response to changes in custom preferences. The big difference between c-commerce and in-house concurrent engineering is that c-commerce requires integrated processes, pervasive information sharing, cooperation, and trust across firms.

The possibilities for information sharing include inventory, sales, demand forecast, order status, product planning, logistics, production schedule, etc., and can be summarised as three types: product information, customer demand and transaction information, and inventory information. Each of these topics will be reviewed in the following:

- *Product information.* Original exchange of product information among the supply chain partners was done by paperwork, such as paper catalogue, fax, etc. The problems caused by this included delays in information sharing and miscommunications among the trading partners. To add the product information into its information systems, a retailer has to re-enter the data, which may or may not come along with the product, manually. Then, keeping the data updated is an even harder task. For example, if some information has been changed since its last release, all the retailers in the industry (if they are lucky enough) have to check the data individually. According to UCCnet, 30 % of data exchanged between suppliers and retailers doesn't match up due to the inefficiencies of manual data entry and convoluted processes (see Figure 2.5 as an example of data synchronisation from [40]).

  This is an enormous problem for the industry, because incorrect data translates into an erroneous understanding of what retailers actually have on their shelves and what suppliers actually have in their warehouses. Faulty data translates directly into huge costs, missed revenues and, often enough, end-user dissatisfaction such as, for example, when shoppers find that heavily advertised products aren't in stock. According to a case study conducted by Vista Technology Group (a CPG software provider), Shaw's (a supermarket chain that has been serving New Englanders for over 140 years) manual, paper-based new item introduction process had no less than 17 steps. This meant a labyrinthine, time-consuming internal process; it also meant that suppliers' product updates — even something as simple as changing the size of a can of tomatoes — had to go through the same manual, error-prone procedure before Shaw's could get the data into its systems. EDI was first introduced for data interchange. Although EDI was originally designed to be a means to process transactions, it has been extended to facilitate sharing of some information like POS and on-hand inventory [10]. However, EDI has its own limitations. In addition, EDI does not verify data accurateness; it just transmits the data — "Garbage in, garbage out".

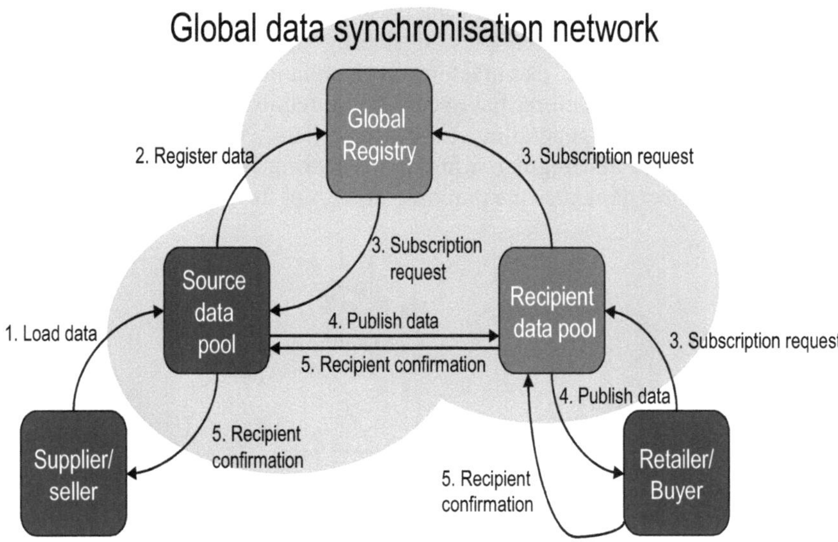

Figure 2.5. Sample supplier-retailer data sinchronization network (adapted from [39])

- *Customer Demand and Transaction Information.* Customer demand and transaction information serves as a critical source of information about future business, and is directly used for demand forecasting, manufacturing schedule, transportation planning, etc. Lee and Whang [11] provide an example of transaction information sharing in Seven-Eleven-Japan's (SEJ). In the SEJ case, POS data are transmitted to SEJ headquarters, wholesalers, and manufacturers to monitor stocking levels, shelf space organisation, merchandising, and new product development. The recent developed Collaborative Forecasting and Replenishment (CFAR) is a new inter-organisational system that enables retailers and manufacturers to forecast demand and schedule production jointly [12].
- *Inventory information.* Including inventory status and inventory decision models, directly affects the amount of orders placed to the immediate upper stream supply chain partners. However, inventory information seems to be more sensitive than customer demand and transaction information (see Figure 2.6), and the trading partners are less willing to share it. For example, manufacturers may not be willing to divulge their true inventory situation or may portray false inventory levels to discourage competitors from producing additional products or building additional capacities and suppliers may use inventory and sales data to get a better bargaining leverage. In practice, sharing of inventory information is implemented in different forms. CRP (Continuous Replenishment Programs) or Vendor-Managed Inventory (VMI) is a practice often employed by two neighboring partners in a supply chain. In a typical CRP relationship, the buyer shares his inventory data with the vendor and asks the vendor to manage his inventory within a

guideline. Wal-Mart's Retail Link program [13] and Apple-Fritz Supplier Hub [11] are good examples of sharing inventory information. VMI system permits the manufacturer to maintain the retailers inventory levels. The manufacturer has access to the retailers inventory data and is responsible for generating purchase orders. The major difference between VMI and regular information sharing is that, under VMI, the manufacturer generates the purchase order, not the retailer.

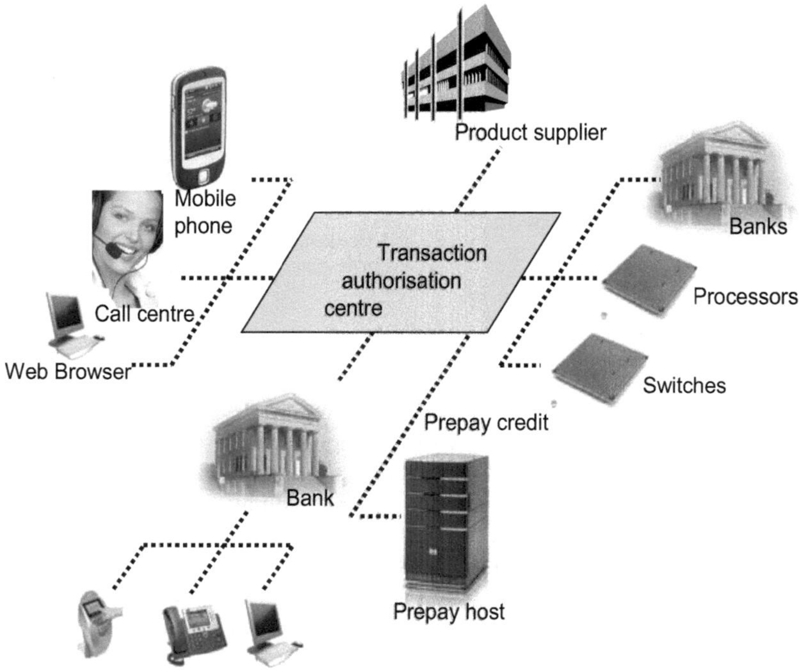

Figure 2.6. Sample system for customer transaction information

2.5 Developing Collaborative Planning Activities

An important effort is needed in terms of both effectiveness and efficiency of the information flows along the chain. As we have seen in the previous section, the information systems are essential to managing a SC, and there is a wide consensus on the idea that the information systems integration is a must [14–18]. Now we will discuss the utilisation of these information systems to improve overall planning activities, which really creates a competitive SC advantage. We shall then review this topic and its latest developments.

In a traditional vision of the supply chain, demand flows up the chain (from each trading partner to its upstream trading partner) and products are moved in the opposite direction (see Figure 2.7). Delay times, distorted demand signals, and poor visibility of exception conditions result in critical information gaps and serious challenges for supply chain managers, including misinformation and, ultimately, mistrust. For example, when partners lose faith in the forecast they receive, they typically respond by building up inventory buffers to guard against demand uncertainty. The disruption that results from dramatic, sudden changes in forecasted demand is amplified as it travels up through the supply chain. This "bullwhip effect" is responsible for much of the inefficiency in supply chains [19–20].

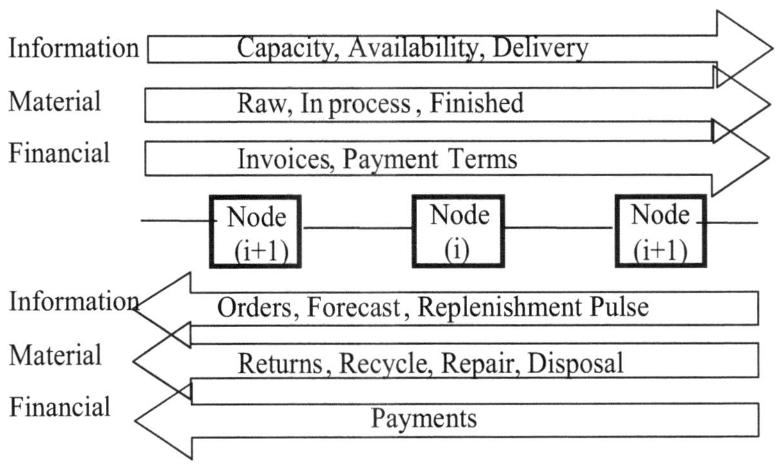

Figure 2.7. Vision of the supply chain flows

The need for certain coordination among the organisations which are participants in a SC should be translated into process and functions integration within these organisations and along the entire SC [6]. Most authors are of the opinion that the benefits of closing the information gap to form collaborative partnerships far outweigh the risk (financial analysis suggests that collaborative planning can lead to inventory reductions of 10% to 50% for each of the supply chain members).

The emergent e-collaboration tools enable the trading partners to exchange business information in supply chain operations, in a structured, agile (in real time), stable and leveraged way [9, 12, 21]. While the collaboration and synchronisation of all SC participants, both within and outside the firm, is now feasible, such supply chain integration needs to be carefully studied in order to improve its implementation. Notice that the term "supply chain automation and collaboration" has gained attention only in recent years, regardless of the fact that various forms of supply chain information exchange systems have been around for over 20 years; for example, Electronic Data Interchange (EDI) and Electronic Funds Transfer (EFT) technologies were first introduced in the late 1970s [22], as we have discussed in the previous section.

Issues involved in supply chain integration improvements have been studied from various perspectives in literature. The reader, for instance, is referred to the following examples:

- Gavirneni *et al.* [18] analysed the benefits of the integration of information flows in a supply chain for a capacitated two-echelon SC;
- Chen *et al.* [20] studied the importance of having access to accurate demand information for the SC upstream members;
- Wikner *et al* [23], Towill *et al* [24], and Chen *et al* [20] have the benefits of integrating the SC and diminishing the demand oscillation transmission along the chain (the bullwhip effect).

Researchers agree that SC planning and control activities need to be considered for a proper SC integration [25] since they have an important impact on the effectiveness and efficiency of the SC.

When considering planning and control activities, the effectiveness of SC integration may depend on the integration process and on the tools used for the integration. This issue deserved attention in the existing literature. For instance, Stevens [16] presented an integration model with four phases:

1. baseline;
2. internal functional integration;
3. integrating supply and demand along the company's own chain;
4. full supply chain integration. Described in terms of reaching a customer-driven supply chain instead of a product-driven one.

Hewitt [26] expanded Stevens' model with a fifth phase that would be dedicated to better administration and re-engineering of the global business processes, pursuing the total effectiveness and efficiency of those processes.

Bowersox [2] also discusses the idea of two types of integration: internal and external. He concluded that the companies need to have a high level of internal integration to be good candidates for the extensive external integration within a supply chain.

By reviewing the practices in the industry under the perspective of supply chain integration, Bowersox found two types of generic integration schemes:

- *The basic integration scheme*, where the SC has developed a set of initiatives and agreements in order to improve connections with customers and suppliers. Under this scheme, benefits are reached through information sharing and common forecast and planning. Such agreements are implemented many times by establishing new venture companies or specific contracts with different members of the supply chain.
- *The advanced integration scheme,* which enlarges the collaboration horizon to reach a more sophisticated dimension. The idea is to integrate the value creation processes with a total end-customer driven orientation. The goal is collaboration to improve competitiveness through a coordinated effort that is, at the same time, feasible in a lean environment (therefore, it results in a reduction in the number of total resources of the supply chain). This advanced integration is normally implemented through profound long-term agreements between companies, and positions the supply chain as an effective competitive unit. Finally, Bowersox suggests that the creation of time and location benefits not

only requires sharing the information to allow suitable business agreements with that purpose, but also requires the existence of a suitable environment for financial transactions.

Another phase model to reach an integrated supply chain: is presented by Scott and Westbrook [27]. They propose three phases:

1. *phase of study*, where everything related to lead times and inventory levels is analysed for potential improvements;
2. *positioning phase*, to identify new opportunities emerging as a consequence of collaboration activities among the members of the chain; and
3. *action phase*, to put previous plans into effect.

Towill *et al.* [24] present an SC integration approach that is similar to that presented by Scott and Westbrook [27]. In their work, Towill *et al.* [24] also use operations management principles to reduce the amplification of the demand signal along the chain when the integration is produced.

Ellram and Cooper [14] identified a set of characteristics that would influence a company's decision to be a part of an integrated supply chain. These characteristics are related to the current level of internal process and functional integration of the company, and with the required level of inter-companies integration for the competition with other SC. Therefore, the importance of those characteristics may differ along the SC integration process [6].

As mentioned above, advanced integration not only requires sharing the information to allow suitable business agreements with that purpose, but also requires the existence of a suitable environment for financial transactions.

The integration of SC financial flows is also becoming a common topic in the literature because of its impact on the entire supply chain performance. Automated freight payment software is available to pre-audit, summarise, batch, and pay carriers by electronic checks on a scheduled basis [28]. There is evidence [29] that the use of information integration in conjunction with buyers' and sellers' banks to transfer funds can improve cash flow and reinforce the "partnering" relationship between the parties in the supply chain. Furthermore, in many supply chains, credit provision is a key factor in supplier choice among distributors and their customers [30]. Suppliers often finance their customers' transactions through the extension of free credit (in Neals' study, only 1% of the distributors charged interest for credit given to their customers, only 5% were charged interest for credit taken, only 12% offered more generous price discounts when customers did not take credit and only 5% received a larger discount when they did not take credit from suppliers).

Clearly, cash flow is affected by the terms of sale, and buying and selling companies often have a different capital cost, which raises the opportunity of improving supply chain performance by having the company with the lowest cost of capital own goods for as long a period as possible [4]. Frequently a financial organisation can provide the "banking function" financing shipments by purchasing those receivables, at a discount, eliminating the seller's extension of credit terms and their incurring of payment delays from letters of credit [31].

2.6 Suppliers Management. Expanding the Purchasing Role

Some industrial sectors, such as hi-tech, face volatility from unpredictable demand and very short product and technology life cycles. Organisations within these sectors develop flexible procurement strategies to deal with this uncertainty.

The numbers of suppliers available, plus a range of tiered contract structures, are critical to meeting the need for flexibility. In such sectors, worldwide capacity for certain parts may be very limited relative to demand at any stage of the commodity's technology life cycle. The global supply is also vulnerable to unexpected events (such as natural disasters, social-political changes, terrorism, and economic disasters) that may create scarcity in worldwide supplies of certain commodity parts.

When products are strategically important for the company, multiple sourcing of strategic parts is used to decrease exposure to potential loss, but in addition companies are now combining supplier contracts types to ensure availability of supply at a competitive cost. This role, creating and managing tiered contract structures for supplier management, is lately becoming a strategic topic, which is necessary to assess the capacity of the organisation for high performance [32].

A strategic part is considered as a part that is critical to product success, with global price and availability driven by external market forces beyond the buyer's control. According to Clark and Fujimoto [33], among other things, organisations need to develop functional specialisation in the area of purchasing strategic parts. This specialisation can of course be shared among many projects running simultaneously, but it is a must for an effective structural design of the organisation as a whole. In this sense, Fujimoto considers that functional specialisation, besides internal integration (inter-functional coordination mechanisms) and external integration (informational consistency between the organisation and the market) are key aspects to take into account.

This expansion of the purchasing role is required to secure an adequate supply in global markets, while protecting profit margins under pressure from global competition. Giunipero and Brand [34] developed a framework describing the stages of the evolution towards supply chain management (SCM) and how procurement would change within that framework. They defined four levels of development of the purchasing role:

1. traditional; emphasizing vendor selection and lowest possible price;
2. partnership/relational; building closer relations with a supplier to reduce total cost and minimize risk in an atmosphere of trust;
3. operational; (material logistics management), coordinating material and information flows to improve quality, inventory levels, and overall cost;
4. strategic; (integrated value added), applying flexible business processes to a given situation, and thereby achieving speed, flexibility, and competitive advantage in the marketplace.

In large multinational companies, the current movement to consolidate supply chain management across business units in geographic areas, and the integration of product units into customer-facing solution businesses by target market, offers new possibilities for strategic sourcing and a contract portfolio.

The common idea is to create consistent relationships between the suppliers of a commodity-type part and the various procurement organisations, locking in competitive prices for the same contractual terms, tracking different product part specifications to a corporate-wide technology strategy [35], etc.

Competitive procurement strategies [36] focus on the buyer's intrinsic bargaining power, which allows buyers to leverage purchasing on a global scale, minimise internal costs, and improve the company's competitive advantage. In this context, global sourcing [37] is a fundamental corporate strategy aimed at maximising the utilisation of worldwide material resources.

2.7 Approaching Markets Differently

The evolution in the way that businesses approach markets has been a frequent literature topic in recent years. For a long time, many companies have mainly focused on their products and processes improvements, trying to develop their technology through creativity and innovation, looking to be in the best market position for every potential customer. However, this sometimes resulted in a poor strategy to attract and retain many customers. Lack of external view and misunderstanding of what the customers really wanted were sometimes the causes of very negative and expensive experiences.

Moreover, in a modern SC scenario, intermediate firms along the CS are also customers in the process; therefore it is also important to ask how the creation of value for intermediate customers influences the behaviour in the channel [4]. Some authors [38] argue that in an SC context, customer success rather than customer satisfaction is the objective of the value-creation process, requiring a firm understanding of what is important to its customers' customer and help immediate customers to deliver that value downstream.

In the hi-tech sector, marketing intelligence and customer knowledge will not only help in the way a business may approach new markets, but will also improve understanding of demand generation and forecast, as a main input of subsequent collaborative planning processes within the entire supply chain. This will ultimately become a key tool to strengthen risk mitigation strategies.

2.8 References

[1] Poirier CC, (1999) Advanced supply chain management. San Francisco: Berret-Koehler Publishers, Inc.
[2] Bowersox DJ, (1997) Integrated supply chain management: A strategic imperative, presented at the Council of Logistics Management 1997 Annual Conference, 5–8 October, Chicago, IL.
[3] Cavinato JL, (1992) A total cost/value model for supply chain competitiveness. Journal of Business Logistics, 13(2): 285–301.

[4] Mentzer JT, (2001). Supply chain management. Thousand Oaks, California: Sage Publications Inc. 306–319.
[5] Chandra C, Grabis J, (2007) Supply Chain Configuration. Concepts Solutions and Applications. New York: Springer.
[6] Cooper MC, Lambert DM, Pagh JD, (1997) Supply chain management: more than a new name for logistics. The International Journal of Logistics Management, 8(1): 1–14.
[7] Rice JB, Hoppe RM, (2001) SC *vs.* SC: The hype and the reality. Supply Chain Management Review, 5(5): 46–54.
[8] Shore B, (2001) Information Sharing in Global Supply Chain Systems. Journal of Global Information Technology Management, 4(3): 27–50.
[9] Gartner Group, (1999). C-Commerce: The new arena for business applications. Business Wire, 16.
[10] Sokol P, (1995) From EDI to Electronic Commerce. New York: McGraw-Hill Inc.
[11] Lee H, Whang W, (1998) Information sharing in a supply chain. Research paper No. 1549, Stanford University.
[12] Raghunathan S, (1999) Interorganisational collaborative forecasting and replenishment systems and supply chain implications. Decision Sciences, 30(4): 1053–1071.
[13] Gill P, Abend J, (1997) Wal-Mart: The supply chain heavyweight champ. Supply Chain Management Review 1(1): 8–16.
[14] Ellram LM, Cooper MC, (1990) Supply chain management, partnership, and the shipper – third party relationship. The International Journal of Logistic Management 1(2): 1–10.
[15] Houlihan JB, (1985) International supply chain management. International Journal of Physical distribution and Materials Management, 15: 22–38.
[16] Stevens GC, (1989) Integrating the supply chain. International Journal of Physical Distribution and Materials Management, 19: 3–8.
[17] Ellram, L. M. 1991. Supply chain management: the industrial organisation perspective. International Journal of Physical Distribution and Logistics Management 21(1): 13–22.
[18] Gavirneni S, Kapuscinski R, Tayur S, (1999) Value of information in capacitated supply chains. From quantitative models for supply chain management. Eds. Magazine MJ, Tayur S, and Ganeshan R, Kluwer: Cambridge.
[19] Lee H, Padmanabhan V, Whang S, (1997) The bullwhip effect in supply chains. Sloan Management Review, 38(3): 93–102.
[20] Chen F, Drezner Z, Ryan JK, Simchi-Levy D, (1999) Quantifying the bullwhip effect in a supply chain: The impact of forecasting, lead times information. Working paper, Northwestern University.
[21] Bauknight DN, (2000) The supply chain future in the e-economy. Supply Chain Management Review, 4 (1): 28–35.
[22] Adam NR, Dogramaci O, Gangopadhyay A, Yesha Y, (1999) Electronic Commerce: Technical, Business, and Legal Issues. Prentice-Hall.
[23] Wikner J, Towill DR, Naim NM, (1991) Smoothing supply chain dynamics. International Journal of Production Economics, 22(3): 231–248.
[24] Towill DR, Naim NM, Wikner J, (1992) Industrial dynamics simulation models in the design of supply chains," International Journal of Physical Distribution and Logistics Management., 22(1): 3–13.
[25] Jones TC, Riley DW, (1985) Using inventory for competitive advantage through supply chain management. International Journal of Physical Distribution and Materials Management 15: 16–26.

[26] Hewitt F, (1994) Supply chain redesign. The International Journal of Logistics Management, 5(2): 1–9.
[27] Scott C, Westbrook R, (1991) New strategic tools for supply chain management. International Journal of Physical Distribution and Logistics Management, 21(1): 23–33.
[28] Cooke JA, (1996) The check in the computer. *Logistic Management*, 32(12): 49–52.
[29] 0rr B, (1996) EDI: Banker's ticket to electronic commerce. *ABA Banking Journal*, 88(5): 64–70.
[30] Neal B, (1994) Springing the distribution credit trap. *Credit Management*. December: 31–35.
[31] Davis K, (1998) Cash forwarding expands business for University Medical Products. *Business Credit*. 100(2): 10–12.
[32] Scott WR, (1987) Organisation: Rational, natural and open systems. 2nd. Edition. Englewood Cliffs, NJ. : Prentice-Hall International Edition.
[33] Clark KB, Fujimoto T, (1991) Product development performance: strategy, organisation and management in the world of auto industry. Harvard Business School. Boston.
[34] Giunipero L, Brand RR, (1996) Purchasing's role in supply chain management. The International Journal of Logistics Management, 7(1): 29–38.
[35] Nellore R, Motwani J, (1999) Procurement commodity structures: issues, lessons and contributions. European Journal of Purchasing and Supply Management, 5: 157–166.
[36] Spekman RE, (1988) A strategic approach to procurement planning. Journal of Purchasing and Materials Management. 25th Anniversary. 4–8.
[37] Arnold U, (1989) Global sourcing- an indispensable element in worldwide competition. *Management International Review*. 29(4): 20.
[38] Fawcett SE, Fawcett SA, (1995) The firm as a value-added system: Integrating logistics, operations, and purchasing. International Journal of Physical Distribution and Logistics Management, 25(3): 24–42.
[39] http://www.uccnet.org/gdsn.html
[40] http://www.uccnet.org

3

Models for SCM Simulation and Analysis

3.1 SCM and Dynamic Simulation

In order to study the SCM issues reviewed in the previous chapter, modern computer simulation methods are perfectly apt. What are the reasons for this? Here are some the features that characterise supply chains as one of the systems best suited for dynamic simulation:

- They are dynamic – that is to say, they display distinctive behaviour that is known to vary through time. Consider, for instance, the dynamics by which the manufacturers adjust their commodity shipments over time on the logistical network, whereas the prices do so over the financial network, or those by which the consumers adjust their consumption amounts based on the prices of the product at the demand markets. There are also interesting dynamics [1] by which the prices adjust over time, etc.
- They are interactive – that is to say, system components interact with one another, and their interaction produces the distinctive behaviour of the system. In modern supply chains, many interesting dynamics are produced by different types of interaction, for instance buyer-supplier interaction, interaction between the user and the provider of information services [2], etc.
- They are complex – that is to say, many objects interact in the system of interest, and their individual dynamics need careful consideration and analysis. The operations performed within a supply chain are a function of a great number of key variables which often seem to have strong interrelationships. The ability of understanding the network as a whole, analysing the interactions between the various components of the integrated system and eventually supplying feedback without de-composing it makes dynamics modelling an ideal tool to model supply chain networks [3].

A review of the traditional supply chain problems studied in dynamics modelling literature shows that these problems are generally related to [3]:

- strategic level decisions: location/allocation decisions, demand planning, distribution channel planning, strategic alliances, new product development, outsourcing, supplier selection, pricing, and network structuring at the strategic level;
- tactical level problems: inventory control, production distribution coordination, order/freight consolidation, material handling, equipment selection and layout design;
- operational level issues: vehicle routing/scheduling, workforce scheduling, record keeping, and packaging, etc.

Different authors recognise MIT Professor Jay W. Forrester as the predecessor of many other studies in the literature related to production- distribution systems modelling. In 1961 Forrester published his book entitled *Industrial Dynamics* [4] where the foundations and major concepts and issues related to the modelling of extended enterprises and supply chains, considering flows of different natures, were highlighted [5]. Indeed, this book is also considered as the seminal work of System Dynamics methodology.

Articles addressing more specific issues related to the dynamic modelling of the supply chains are published mainly since the late 1990s [5], and they address, for instance, issues such as:

- The dynamic assessment of the performance of different SC nodes and for certain scenarios [6, 7]. Studying the impact of possible overload of production units, inventory shortages, or the bullwhip effect would fall under this category of issues. Dynamic models help to understand the supply chain dynamics better and serve as a decision support tool to determine the impact of possible allocation strategies for human and technological resources [8]. In these cases, simulation has become a powerful approach to assess and design global SC strategies.
- The study of both the flexibility and the reactivity of the supply chains to unexpected events [9] (e.g. logistic perturbations, raw material shortages, etc.).
- Assessment of the efficiency of the logistic system using dynamic models has led to intensive publications addressing modelling methods for supply chains and for networks of plants. Object oriented modelling has been suggested by Changchien and Shen [10], and Biswas and Narahari [11] to model SC, for evaluating and analysing reengineering proposals, and for decision support purposes.
- Operational and financial assessment of internet based applications to improve SCM [12, 13].

Regarding the dynamic simulation techniques used in the published papers we can appreciate in literature different approaches, for instance:

- Classical discrete event simulation packages have been used to model the SC and to provide animation capabilities (e.g. Arena®, Witness or

PROSIM). The use of other well known simulation tools, such as Taylor II and Automod is reported in [7]. Also dedicated environments based on simulation tools as ARENA® or ExtendSim® are showing up as a consequence of the growing interest in SC simulation.

- Because of the uncertainty of the production environment, fuzzy sets theory has been incorporated in the modelling approach [14].
- Agent–based simulation has also been recently used as a tool providing excellent representation of supply chain operations and concurrent activities, allowing for very detailed operational data to be gathered [15].

3.2 Continuous Time Simulation Models for SCM

Supply chain simulation studies have been, as noted above, frequently based on a discrete event worldview [16]. As previously done for classical manufacturing system simulations – modelling the flow of individual products through a set of production resources (e.g. machines, operators, and automated guided vehicles), waiting in queues if necessary – at the supply chain level the flow of batches of products (instead of products) are modelled. The flow is now between production units or work centres (instead of machines), and this flow is accumulated in inventories and flow from one unit to another using logistic resources (e.g. trucks). In certain publications a combined discrete event continuous approach is also suggested. Researchers argue that "SCs are neither completely discrete nor continuous". Although sometimes the SC ''continuous nature'' is not obvious, important parts of the dynamic behaviour of the SC can be described in a relevant way using equations, especially when one is concerned with strategic activity levels [16].

The use of continuous time simulation and more specifically System Dynamics in production systems has been discussed in several research papers (e.g. [17–19]). In 1989 John Sterman [20] presented a generic model of a stock management system using System Dynamics that has been then applicable to many different SC scenarios, including raw material ordering, production control, or at a macroeconomic level, the control of the stock of money. The model consists of two parts, the physical stock and flow structure of the system, and the decision rules used to control the system.

Sterman work shows clearly potential application of System Dynamics to SCM. It shows how misperception of feedback loops can be responsible for poor performance in dynamic SC decision-making. His model is extremely powerful when combined with the use of the Beer Game [21]. This game allows the staging of an experiment in managing a simulated industrial production and distribution system. The Beer Game presents a multi-echelon production distribution system, containing multiple actors, non-linearities, feedbacks and time delays throughout the supply line. The players are advised to minimise costs by managing their inventories under uncertain demands and unknown delivery lags. During the course of a simulation run, the system exhibits oscillations and amplification of variable values. Sterman [20] is able to reproduce behaviour of the system by modelling the decision-making process with locally rational heuristics in the form of an anchoring and adjustment policy.

Other interesting contributions related to continuous time simulation and SCM are presented by Towill [22, 23]. His research on supply chain re-design by time compression strategies helps to explain, for instance, the impact of these strategies in terms of SC response velocity, effectiveness and efficiency when there are SC market changes. Towill proposes, for this instance, that time compression strategies based on simulation allow one to predict supply chain performance improvements.

At present, many models are being built using a systems dynamics (SD) approach [3], and for different industries [24]. The operations performed within a supply chain are a function of a great number of key variables which often seem to have strong interrelationships. The ability to understand the network as a whole, analysing the interactions between the various components of the integrated system and eventually supplying feedback without de-composing it make systems dynamics an ideal methodology for modelling supply chain networks. The objective of many papers is to model the operation of the supply chain network under study and to obtain a true reflection of its behaviour. The modelling framework is also used to study the performance of the system under different scenarios concerning commonly addressed real-life operational conditions. Modelling efforts are also focused on measuring the supply chain system performance in terms of key metrics such as inventory, WIP levels, backlogged orders and customer satisfaction at all SC nodes.

3.3 Classifying Hi-tech SC Dynamic Models in this Book

3.3.1 Overview

Hi-tech industry and customers are moving towards a new model of computing, a model based on industry standard, market unifying technologies and architectures. This new model requires from hi-tech companies an offering of, not just great products, but great solutions. The organisations need to become a more customer-focused organisation. In order to do so, many corporations are migrating from product divisions to a "front-back" organisational model. This model designs the organisation considering two units called "front end" and a "back end", as follows:

- The "front end" units deal with units handling sales and marketing, and are organised according to customer type. These units are able to offer specific integrated solutions to customers.
- The "back end" units deal with research, development and elements of manufacturing. They are are organised by product or technology type, and they are able provide the modular elements to be combined into solutions.

This structure was suggested in the early 1970s, and hi-tech companies such as Xerox, Motorola, IBM, Lucent and lately HP have experimented with it. Commercial and investment banks often also have structures resembling a front-back model, with customer groups specialised by industry (retail, telecommunications, transport, etc.) and product groups by instrument (cash management, derivatives, debt securities, etc.).

However, front-back structures are notoriously difficult to make work. The problem is how to organise one manner at the front, one manner at the back, and somehow link or integrate the two together.

The models presented in this book attempt to help in this process. We try to show how solutions to problems within this area can be found through the use of appropriate dynamic simulation models.

We will concentrate on hi-tech supply chains and network problems inside a front-back organisational model. These problems are related to many different topics of management science such as marketing, operations, financial and risk management, etc.

We will face special challenges to find an appropriate solution using our models and experience how the need for an interdisciplinary approach in dynamic modelling is increasing.

We have grouped the problems, and therefore the models into three areas as we will explain in the following sections.

3.3.2 Front-end Dynamics Modelling

In this, the second part of the book, we will review and propose dynamic modelling options to connect customer value to business targets. This is done by explaining how to characterise target market by formalising what are often informal but deeply held beliefs about what drives customers' purchase decisions.

We will explain how dynamic models may help to connect planned investments to expected improvements in the customer's perception of the product's critical attributes and thus increase sales, revenue, and market share. With the same effort we can improve our customer demand forecast, achieving a much better input for subsequent integrated supply chain planning models.

Our models are based on a general overall high level model presented in Figure 3.1, which is a representation of numerous planning team dialogues about the way a business grows when it offers a valuable product to an existing market. The diagram links operational investment, conditioned by policy, to business revenue growth over a financial year. In this way, financial constraints are introduced into the model. Obviously, the higher the growth at a reasonable margin, the greater the level of investments that are available for the following year. Of course, this simplified diagram does not show all the exogenous and endogenous factors that condition results over time, and that are included in this model for a valid simulation.

For a myriad of reasons business planners know that, over time, it takes more dollars of investment to grow or sustain share (this, of course, does not apply to all cases, e.g. if a big rival has failed, the firm may be able to grow or sustain share with less expenditure). The model indeed shows diminishing returns over time, depending on a number of factors. Most importantly, the model clearly shows why "doing nothing" is almost never a wise decision for a hi-tech business, and helps a business that has enjoyed great success in the past to act aggressively to protect its position for continued profit and growth.

Incremental investments are represented in this model as completely variable, even though volume ramps up or down would surely affect the return on fixed costs. We do not include a fixed costs component simply because none of the financial or strategic planners among the companies we worked with have done so. Industry practice is to build fixed costs into overhead rates as part of labour, material, and overhead in internal part costs, or priced into purchased parts, and are not visible to our clients nor used by them when they evaluate and compare business plans.

The allowable change in spending level corresponds to an expected changed value of specific attributes. Note that the investment cycle is a consequence of corporate policy and regulated periods to report results and commit resources, where external economic cycles and market occur at their own pace. The model recognises the delays between a change in spending and a resulting improvement in customer value and sales growth.

Business planners further attempt to group their customers in segments within the target market, according to the relative importance the buyers place on one or another of the attributes that drive their market overall.

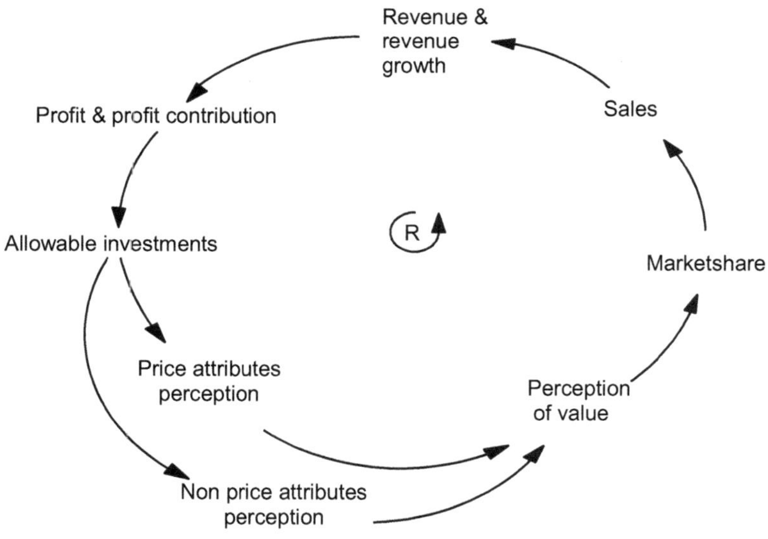

Figure 3.1. General model overview (original team design)

In a scenario, investing to improve product attributes drives positive change in customer perception, which are assumed by business planners to drive each competitor's share in each segment of the overall market, and of course the related financial results.

The financial model is set by reporting rules, the investment model by budget and targeting practices, and a value index computed from quality relative to price has gained wide acceptance and general industry use. In this work, we will compute the value index in a manner that takes advantage of the capabilities of System Dynamics for the benefit of fast moving hi-tech industries.

In terms of confirmation and validation, the general model structure that we present in Figure 3.1 was synthesised and refined with commercial and consumer business managers, systems analysts, critical part contract managers, financial executives and experts in hi-tech workforce collaboration.

3.3.3 Modelling Back-end Issues in SCM

In researching how to manage and value a portfolio of supplier contracts, we will apply dynamic modelling to model the strategic parts procurement system. This work will attempt to illuminate the dynamics of the procurement process, and to assess the value of a contract portfolio within this process. We will define, characterise and simulate different generic types of supplier contracts to accomplish varying degrees of security and flexibility. We will then simulate a portfolio of these contracts applied to secure a single part, with the purpose of extending and refining portfolio valuation. We will focus our attention on business dynamics based on current best practices in portfolio management, as evidenced by leaders in volatile technology businesses.

As mentioned in the Preface, this part of the book also deals with manufacturing issues and problems that can be explored by using this methodology. For instance, we will use the dynamic simulation technique for the assessment of alternative scheduling policies that could be implemented dynamically on the shop floor. Policies considered will be based on the manufacturing equipment status or on several operating conditions of the production flow.

As mentioned in Section 3.2 it will be very interesting for the reader to appreciate how we can use this methodology in manufacturing and what its benefits are compared to other classical tools such as LP models.

A strong point of this part is the process to produce simple rules to guide operators' decisions in dynamic scenarios. We have the experience that this approach is far more valuable than a fixed set of times and tasks to perform that do not account for the changing manufacturing environment.

3.3.4 Modelling Integration Issues in SCM

Supply chain management (SCM) requires the coordination of the information, material, and financial flows along different nodes of the supply chain. Therefore, by considering each of these different flows, we can develop different models of a supply chain.

A part of the book (Part IV) is devoted to review and discuss the operational and financial effectiveness of existing virtual tools used in supply chain integration. This is done by explaining how dynamic modelling may help to obtain a comprehensive supply chain integration model.

This modelling effort can be used for the analysis of the effectiveness of various levels of operational and financial integration, as well as for the assessment of the importance of the sequence, in which virtual collaboration tools are adopted, in supply chain integration. The review of existing SC integration literature reveals that there is a need for comprehensive SC integration models. Therefore, the purpose of this part of the work is to develop and evaluate a comprehensive supply chain model that can be used to determine the operational and financial benefits of various levels of supply chain integration using e-collaboration tools. Such an SC model would also enable us to analyse the impact of partial integration efforts.

Some models in this part will also show how to integrate the supply chain planning and control using manufacturing techniques "virtually". Manufacturing rules for materials control can then be applied to supply chain nodes instead of to production stations or equipment.

Finally, some dynamic models for SC integration that are considered here deal with "soft issues" such as human resources integration. A dynamic simulation approach to analyse the cultural diversity integration in modern organisations is presented. Models in this area deal with the so-called "soft variables" and the reader may experience how these dynamic models can provide a balanced perspective to handle both hard and soft system-based problems. At this point, an important question is the validation of the model which incorporates soft variables. The soft variables cannot, by definition, be measured directly and objectively. These are measured by quasi-quantitative methods influenced by systematic and random measurement errors and the structure of relationships among these variables is often unclear.

3.4 References

[1] Nagurney A, Ke K, Cruz, Hancock, Southworth F, (2002) Dynamics of supply chains: A multilevel (Logistical/ Informational/ Financial) network perspective. Environment & Planning, 29: 795–818.

[2] Shee DY, Tang T, Tzeng G-H, (2000) Modelling the supply-demand interaction in electronic commerce: A bi-level programming approach. Journal of Electronic Commerce Research, 1(2): 79–93.

[3] Özbayrak M, Papadopoulou TC, Akgun M, (2007) Systems Dynamics modelling of a manufacturing supply chain system. Simulation Modelling Practice and Theory, 15(10): 1338–1355.

[4] Forrester JW, (1961) Industrial Dynamics. Productivity Press, 1961.

[5] Holweg M, Bicheno J, (2002) Supply chain simulation. A tool for education, enhancement and endeavour, International Journal of Production Economics 78: 163–175.

[6] Petrovic D, Roy R, Petrovic R, (1998) Modelling and simulation of a supply chain in an uncertain environment. European Journal of Operational Research, 109: 299–309.

[7] Terzi S, Cavalieri S, (2004) Simulation in the supply chain context: A survey. Computers in Industry, 53: 3–16.

[8] Pearson DW, Albert P, Besombes B, Boudarel MR, Marcon E, Mnemoi G, (2002) Modelling enterprise networks: A master equation approach. European Journal of Operational Research, 138 (3): 663–670.
[9] Wilson MC, (2007) The impact of transportation disruptions on supply chain performance. Transportation Research Part E. Logistics and Transportation Review, 43(4): 295–320.
[10] Changchien W, Shen HY, (2002) Supply chain reengineering using a core process analysis matrix and object-oriented simulation. European Journal of Operational Research 153: 704–726.
[11] Biswas S, Narahari Y, (2004) Object oriented modelling and decision support for supply chains. European Journal of Operational Research 153 :704–726.
[12] Crespo Marquez A, Bianchi C, Gupta JND, (2004) Operational and financial effectiveness of e-collaboration tools in supply chain integration. European Journal of Operations Research, 159(2).348–363.
[13] Crespo Marquez A, Rubiano O, Framinan JM, (2001) Benefits of the Internet for the supply chain management. A Characterisation and simulation study. International Journal of Agile Manufacturing. Special issue on information systems and agile manufacturing. 4(2): 25–42.
[14] Petrovic D, (2001) Simulation study of supply chains behaviour and performance in an uncertain environment. International Journal of Production Economics, 71: 429–438.
[15] Chatfielda DC, Hayyab JC, HarrisoncTP, (2007) A multi-formalism architecture for agent-based, order-centric supply chain simulation. Simulation Modelling Practice and Theory, 15(2) 153–174.
[16] Lee YH, Cho MK, Kim SJ, Kim YB, (2002) Supply chain simulation with discrete continuous combined modelling, Computers and Industrial Engineering 43 (1–2) (2002) 375–392.
[17] Pierreval H, Bruniaux R, Caux C, (2003) A continuous simulation approach for supply chains in the automotive industry. Simulation Modelling Practice and Theory, 15 (2): 185–198.
[18] Edghill J, Towill D, (1989) The use of System Dynamics in manufacturing system engineering. Transactions of the Institute of Measurement and Control, 11(4): 208–216.
[19] Thiel D, (1996) Analysis of the behaviour of production systems using continuous simulation, International Journal of Production Research, 34 (11): 3227–3251.
[20] Sterman J D, (1989) Modelling Managerial Behaviour: Misperceptions of feedback in a dynamic decision making experiment. Management Science, 35(3): 321–339.
[21] Sterman JD, (1984) Instructions for Running the Beer Distribution Game: MIT System Dynamics Group.
[22] Towill DR, (1996). Industrial dynamics modelling of supply chains. Logistics Information Management, 9(4): 43–56.
[23] Towill DR, (1996) Time compression and supply chain management – a guided tour. Supply Chain Management 1 (1):15–27.
[24] Hafeez K, Griffiths M, Griffiths J, Naim MM, (1996) Systems design of a two-echelon steel industry supply chain. International Journal of Production Economics, 45(1–3): 121–130.

Part II

Modelling Front-end Issues in SCM

4

Understanding a Customer's Decision to Buy

4.1 Selecting Potential Markets

Companies want to be able to determine what the potential market is for their product or service, as well as the best ways to reach potential consumers.

In order to know the potential market they must identify the characteristics of individuals likely to be interested in a particular product or service, establish how many such individuals there are, as well as study how these people behave and respond to particular advertising approaches.

A group of people or organisations sharing one or more characteristics causing them to have similar product needs is called a market segment. An enterprise can achieve dramatic success in the marketplace by finding a new way to divide customers into groups whose needs differ distinctly. This allows the enterprise to offer each segment a benefit package that is unmistakably tailored to meet that segment's uniquely different needs [1]. A true market segment meets all of the following criteria:

- it is distinct from other segments (heterogeneity across segments);
- it is homogeneous within the segment (exhibits common attributes);
- it responds in a similar manner to a market stimulus, and it can be reached by a market intervention.

Market segmentation is the process of partitioning markets into these groups of potential customers with similar needs or characteristics who are likely to exhibit similar purchase behaviour.

Segmentation is generally conducted using demographic, geographic, attitudinal or behavioural data. If each segment is fairly homogeneous in its needs and attitudes, it is likely to respond similarly to a given marketing strategy. That is, each segment is likely to have similar feelings and ideas about a marketing mix (comprising a given product or service, sold at a given price, and distributed and promoted in a certain manner).

The segmentation process in itself consists of segment identification, segment characterisation, segment evaluation and target segment selection. This process can allow an organisation to concentrate its limited resources on the greatest opportunities to increase sales, customer satisfaction and achieve a sustainable competitive advantage. However, a fundamental issue needing rigorous attention is that customers' needs are dynamic and can induce segment instability [2]. Therefore customer value change also has to be deeply explored to avoid serious mistakes when assessing customer segments.

The segmentation process will be critical for our business planning modelling. In the following section we will present some examples with the type of information that the modeller may find and use to model market segments.

4.2 A Case Study for Market Segmentation

This could be an example of a presentation of a segmentation study conducted by a large flat panel display producer "FPD Co.". Several consultants visited and interviewed many of their potential customers; at the end of the day the company was able to define three user application segments where their product could be most compelling. At the same time, the company was evaluating potential problems arising in those segments as a result of the possible need for a periodic re-configuration of the product once it was sold to the customer. Let us see an example of how those segments were characterised. The segments names were:

- data collector;
- data searcher;
- CAD & Creative.

In order to proceed with the Characterisation of these segments the team had to find out what these segment customers did, what type of workers they were, which product in the company they would find compelling, what type of problem they could foresee when buying the product, etc. In Tables 4.1, 4.2 and 4.3 we present a sample report answering the previous questions.

In Figure 4.1, the data collector column is divided into eight sectors according to the following type of workers (top → down):

- adjustment clerks;
- bill collectors;
- insurance policy processing, insurance claims clerks;
- new accounts clerks;
- billing clerks;
- order entry clerks;
- telephone operators; and;
- material dispatchers.

The data searchers column is divided into five sectors corresponding to the following type of workers (top → down):

- financial professionals;
- securities & financial services sales representatives;

- loan clerks and credit authorizers,
- cs & systems analysts; and
- computer programmers and operators.

Similarly, CAD and creative segment column is divided into six sectors corresponding to the following type of workers (top → down):

- architects;
- designers;
- drafters;
- civil engineers;
- mechanical engineers; and
- aerospace engineers.

At the same time, information about the number of workers per segment (segments size) was obtained and it is presented in Figure 4.1.

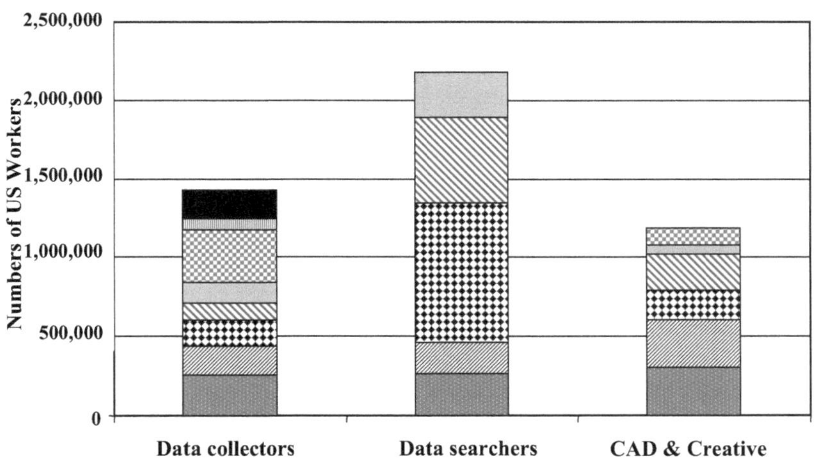

Figure 4.1. Number of workers per segment. Source = 1999 US Occupational Outlook Handbook (Bureau of Labor Statistics 1998 data)

Some other findings of the market segmentation team related to the industry sectors of the above-mentioned workers. In this regard, the following information was gathered:

- Data collectors were found primarily in three industries: banking & finance, insurance, wholesale/retail.
- Data searchers were found primarily in two industries: financial, computer and data processing services.
- CAD and creative were found in small businesses, large engineering or manufacturing firms, and government. Graphic artists are employed by business services, advertising and design firms, or are self-employed.

Table 4.1. Characterisation of the data collector segment

What they do	Gather, edit, manipulate, or enter data into large data base or industry application
Types of workers	Claims & insurance processing, collections/accounting departments, customer service centres, help desks, order entry & telemarketing, phone operators, material dispatchers
What's compelling	Larger screen is affordable (relative to the 19" they possess), a single person can install it, and fewer mistakes since users can see what they're doing
What's a problem	The possible need for periodic re-configuration of the screen causes IT people to worry, calls to their departments, and time spent fixing problems they don't worry about currently
Customer example	Product warranty claims processors

Table 4.2. Characterisation of the data searcher segment

What they do	Research and retrieve, track, read, and file information that they then act on.
Types of workers	Lawyers & paralegals, financial professionals, stock brokers & traders, loan clerks investment bankers, computer scientists, systems analysts, programmers, librarians
What's compelling	Larger screen means workers are more productive (less scrolling in spreadsheets, less switching from application to application). Product is affordable relative to flat panel alternatives, but fits in a small space relative to CRT alternatives
What's a problem	Besides re-configuration for IT people, another problem is mission critical or time sensitive information that could "not be seen"
Customer example	Bank of America treasury department

Table 4.3. Characterisation of the CAD and creative segment

What they do	CAD: create drawings & designs Graphic creative person: create sales material, brochures, newsletters
Types of workers	CAD: draftsmen & architects, designers, civil, aerospace, and mechanical engineers. Graphic creative workers and artist
What's compelling	A larger screen that fits on a desk (relative to multiple 21" screens currently available). It's affordable relative to a flat panel. Colour matching is important (full colour pixels)
What's a problem	They are very "picky" about line shapes, may dislike square pixels, and are more likely to use higher resolutions
Customer example	Engineering company

However, segmentations are only useful if they can be applied. Target segments need to be analytically identified, and after that, further analytic work may answer other strategic questions for each segment, for instance:

- What factors (drivers) impact the outcome of interest, such as purchase intent, intent to prescribe, or intent to use?
- Which key drivers are important?
- What is the market potential?

The market analysis team was told to select segments offering greatest opportunities for the company (see segment map in Table 4.4). This was the result concerning the segments to target and the reason why they were selected:

1. Target large companies because:
   - they are easier to reach than small companies with a focused market entry approach;
   - they have evaluation cycles allowing us to explain our value (not just specifications);
   - however they have a problem with the re-configuration (lots of users, potential calls to IT, total cost of ownership, hassle factor).
2. Target the data entry/knowledge worker segments because:
   - there are many of these types of workers found in large companies in a few industries (insurance, banking, retail);
   - they are likely place high value on size and weight;
   - they aren't as "picky" about lines and pixel shapes as the other two segments.

Finally, the team declared that the next steps of the market analysis process would be:

1. Concept testing to validate assumptions:
   - recruited from all three segments in separate focus groups, talking to both MIS and end-users;
   - test alternative product re-configuration scenarios.
2. Follow up with targeted customer visits in selected segments & industries:
   - understand "a day in life" usage;
   - determine the product purchase process;
   - determine needed sales, channel, and support structures;
   - refine market sizing and unit forecast.

Table 4.4. Sample segment map for case study segments

Segment map (☺ Data collector ☻ Data searcher ♣ CAD & creative)						
Compelling Reason to Buy	Smaller depth ☺ ☻	Lighter Weight ☺ ☻	Bigger screen ☺ ☻	Better image quality ♣	Brighter colours ♣	Image - cool technology
PC location	Home	Small biz ♣	Medium/ large biz ☺ ☻			
Decision maker	Head of house	Biz owner ♣	Dept. mgr.	MIS ☺ ☻	End user	
Buying to…	1st time PC	Replace 15’	Replace 17 or 19’’CRT ☺ ☻	Replace 21” ♣	Instead of flat panel ☺	
Primary PC application	Spread-sheets ☻	Internet Research ☻	Word Processing	Graphics & Images	Design & Graphic Arts ♣	Data base ☺ ☻
Type of document	Present. & memos ☺	Reports & proposals ☺ ☻	Sales materials ♣	Manuals & booklets ♣	Drawings & maps ♣	E-forms ☺
Buying POV	Buy with a PC ☺ ☻♣	Never upgrade	Upgrade to fit budget	Upgrade every 3 years		
Brand POV	Whatever brand	Brand loyal to PC	A name I know & trust ☺ ☻♣			
Where they buy	Retail ♣	Mail order	Value add. reseller ♣	Computer dealer ☺ ☻♣	Direct	Internet
Risk disposition	Early Adopter	Pragmatist ☺ ☻♣	Laggard			
Price sensitivity	Spend no more than CRT	Want more for less	Pay a little more for a lot more ☺ ☻♣	Want less for less		
Key benefit	It fits in a small work space	Easier to install & move	Users can see-fewer mistakes ☺	Productivi-ty gain, less scrolling ☻	Better lines, colours ♣	

4.3 The Monitor Purchase Process. A Case Study

As "FPD Co." began its focused entry strategy in the year 2000, it became vital that the monitor purchase decision process could be understood within the target segments. However in "FPD Co." there was little reported with regard to the FPD monitor purchase process directly. Therefore, the team had to look into the purchase decision process for other similar hi-tech products within the overall segment to see if they could give some insight into how monitors were purchased. In this case study this was a key issue.

It was appreciated that the purchase decision process varied by company, company size, site size, and number of products installed.

Qualitatively companies could be separated into two distinct groups: strategic or tactical (see Table 4.5). For the purpose of this study these companies were characterised as follows:

- *Strategic organisations*: these are organisations that are more proactive in their approach to monitors and IT spending with a long-term focus. They will typically have a minimum standard specification for PCs and monitors. Monitor prices are usually visible by either a separate RFP for monitors or monitors split out on a single RFP for PCs and monitors. They perceive this as a better way to purchase and perceive greater flexibility in selecting a preferred brand.

  Influence on the brand decision ranged from IT staff to procurement to management. The reseller has little or no influence in a strategic organisation where more often the monitor vendor has the greater influence through specifications and Marcom materials.

  Monitor brand is important to strategic organisations where the monitor is typically evaluated internally and a reseller is approached that can supply the desired monitor brand.
- *Tactical organisations*: a tactical organisation has a short term reactive approach to IT spending and acquisitions. They typically replace monitors when necessary. Monitor pricing is typically invisible to these organisations that typically take the monitor bundled with the PC and perceive this to be the least expensive way to purchase.

  The IT manager will have the most influence on which brand is purchased or recommended and in many cases will have autonomous decision making power with little internal support to evaluate brands.

  In a tactical organisation the reseller has a greater influence on what brand is purchased. Typically the tactical organisation relies upon the reseller having confidence in their own knowledge and advice.

Quantitatively, a recent study of purchase decision process in the corporate sector revealed some interesting insights to the overall purchase process.

Intuitively one would expect that the majority of large corporations and some medium companies are either required or encouraged to purchase new equipment only after some qualification testing and then only if the brand or model is on an approved list. This list is often referred to as the 'standards list'.

In the mentioned study, of those companies (70%) that are ether required or encouraged to purchase from a standards list, 66% have a formal requirement (see also Table 4.6). When the use of a standards list is a formal requirement, a majority have a centralised process whereby the list is designed by a group or committee.

The evaluation of new equipment is an important step in the creation of the standards list. Of the corporations, 57% indicated that they conduct internal qualifications tests. Of those that conduct internal tests, 64% indicated that the testing lasts 2 or more weeks with only 35% having tests that last 1 week or less.

Table 4.5. Characterisation of companies according to their purchase decision process

	Strategic	Tactical
Focus	Long term	Short term
Approach	Proactive	Reactive
Monitor replacement strategy	Planned as transitions take place	Replace as necessary
Purchase process	Formal	Less formal to none
Decision maker	Evaluation team	IT manager
Visibility of monitor price	Usually visible – separate RFQ	Usually invisible – purchased as system
Vendor influence	Strong	Weak
Reseller influence	Weak	Strong
Brand importance	Product brand important	Reseller brand important
Evaluation process	Formal	Little or none

A common question in a purchase process study is "Who is the new equipment purchased for?":

- About half (52%) said new equipment is purchased for established employees whose old equipment goes to another employee.
- One quarter said they usually purchase for new employees.

- A figure of23% said they purchase new equipment for established employees whose equipment is disposed of somehow. There was no separate mention of monitor purchase.

When new products with new technology are introduced there seems to be some difference as to whether the organisation continues to purchase the products with the same technology or starts to purchase products with the new technology very soon after its introduction.

This was split more along the lines of brand preference than any other internal procedure. Significantly, most of those who prefer certain brands said their site continued to purchase the same technology products.

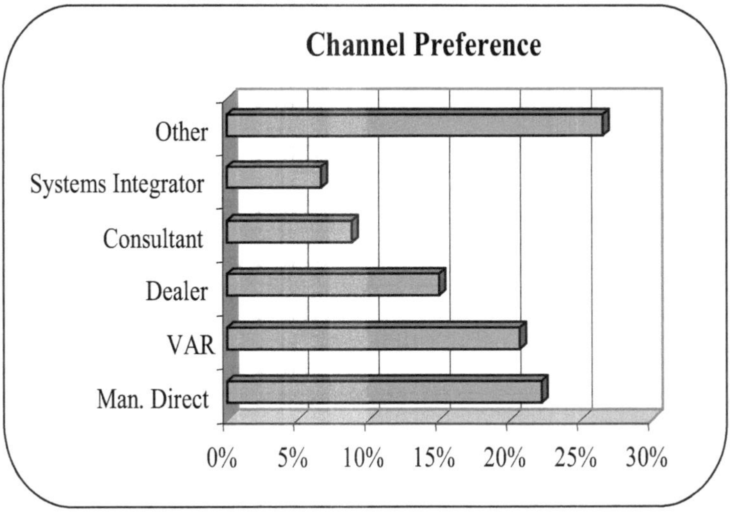

Figure 4.2. Overall results for channel preference

Channel preference was also split along brand preference. Those who prefer Dell, for instance, were likely to buy direct from the manufacturer. Those preferring Compaq or HP were more likely to purchase from a value-added-reseller. The overall channel preference found is shown in Figure 4.2. Product channel preference was found to differ substantively among product type. For instance, in the case of printers most large companies prefer corporate resellers as their printer source. However, there seems to be a growing number of large companies asking to purchase printers directly from the manufacturer.

As can be seen in Figure 4.3, the reasons for not buying direct varied with the most popular reasons revolving around better service and price. The 'other' category included remarks like, "my preferred company does not sell direct, it's company policy, a contract, long standing relationships, or leasing."

The purchase process in the target segment will pose some challenges as the company moved forward with our focused entry strategy. To reach the target customer the company had to penetrate the purchase decision process that in most cases was a formal requirement, dependent on internal qualification testing lasting more than two weeks. Those companies not buying from a direct source were more likely to purchase from a VAR (Value Added Reseller) than a computer dealer. Table 4.6 represents the general purchase decision process for companies that have more than 250 employees at their site, have more than 100 PCs installed and have IT departments.

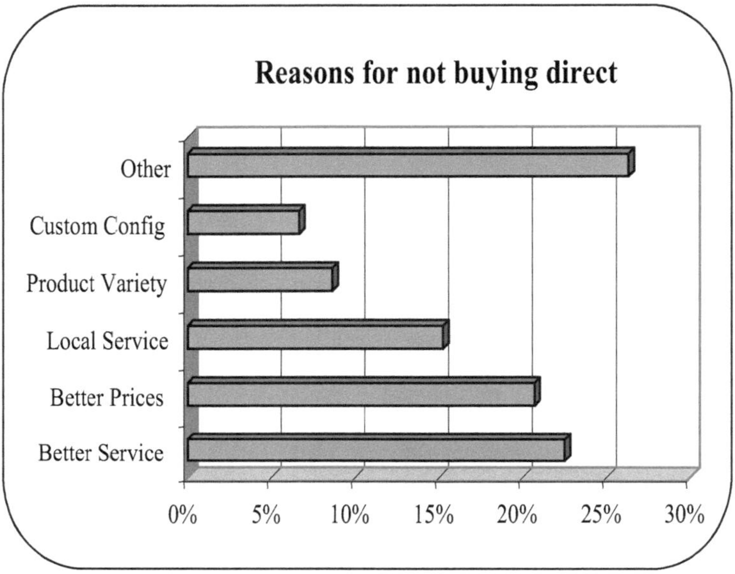

Figure 4.3. Reasons for not buying direct

As "FPD Co." embarked on executing the focused entry strategy for FPD products the company decided to target those companies whose IT purchase decision process was more strategic than tactical, those companies who have a more formal and more centralised process.

This decision would allow "FPD Co." to have the product evaluated on its merits and its ability to satisfy the needs of discreet user segments in those companies. It would also allow strategic positioning against most convenient monitors (CRTs at that time).

From this, the company could also derive its channel strategy. It would follow that in order to get their products into the evaluation process of these companies they would need:

- A channel that could provide a 'directed' sale. Meaning a channel that has an outbound sales force, a 'high-touch' component to their sales model, and could penetrate the segments chosen.
- They would also want to choose a channel partner that would be willing to let them participate in the sales process and let them develop a relationship with some of the purchasing companies.

  The companys' first product would probably not be an "off-the-shelf" sale and therefore they would not be targeting the retail channels used by many other company products and divisions until the technology and the products would be able to create the necessary pull to be sold in that format.

Table 4.6. Purchase decision process for companies that have more than 250 employees at their site, have more than 100 PCs installed and have IT departments

	Percent
Use a standards list	70%
Of those using standards list, percent of companies that are required to purchase from the list	66%
Of those using standards list, percent of companies with a centralised process	58%
Price as highest influencer	32%
Value as highest influencer	47%
Buying criteria importance out of 5	Compatible with existing 4.5 Quality of tech support 4.1 Ease of maintenance 4.1 Purchase price 4.0 Cost of ownership 3.7

4.4 Concluding Remarks

In this chapter the reader has been able to realise the vital importance of a proper market segmentation process and study in order to target the most valuable customers properly. Also understanding the purchasing process, the process of entry, the one required to figure out how to "reach" those customers in the target segments, is critical when approaching markets. Marketing intelligence teams working with dynamic modelling analysts may provide extremely high value to these processes as we will see later, capturing the knowledge of the organisations in this volatile world.

4.5 References

[1] Cleland AS, Bruno AV, (1996) The Market Value Process. San Francisco: Jossey-Bass Publishers.

[2] Blocker CP, Flint DJ, (2007) Customer segments as moving targets: Integrating customer value dynamism into segment instability logic. Industrial Marketing Management. 36(6): 810–822

5

Understanding Financial Implications of Strategy

5.1 Overview

In order to understand financial implications of a certain market strategy it is necessary to analyse the revenue stream that is generated (the way a company makes money, the company's income) besides the existing strategy cost structure (monetary consequences of the means employed in the current business model).

We will now discuss these topics and will once again present a case study for the reader to appreciate the type of information that the modeller may find and use when modelling financial aspects of the supply chain.

5.2 The Price as Source of Revenue Stream

5.2.1 Characterising Pricing Options

Setting the price of a product is a critical decision for business success. The product and its price are the linkages between the buyer and the business. We can therefore say that the price is the source of the revenue stream, and the pricing policy has an enormous impact on it. For instance, a study based on a sample of Fortune 500 companies [1] has shown how a 5% increase in average selling price increases earnings before interest and taxes (EBIT) by 22% on average, compared with the increase of 12% and 10% for a corresponding increase in turnover and reduction in costs of goods sold, respectively.

Pricing has however, received little academic investigation. Not only managers but also academics have shown little interest in the subject of pricing: Publications on this subject are nowhere near as numerous as publications on other classical marketing instruments. Even marketing scholars have devoted little effort to pricing theory and practice: An empirical study revealed that less than 2% of all articles published in major marketing journals cover the subject of pricing [2].

Of course, as a general rule, the minimum price to charge for a product is the cost of producing and marketing it and the maximum is what the buyer is willing to pay for it. Therefore the product price is set somewhere within this range. At the same time, the price of a product, when offered for sale by competitors, will influence where we set the price within this range.

General pricing options and factors influencing the pricing decision making process can be classified as follows:

- *Cost plus profit pricing.* The cost plus profit approach involves adding a predetermined profit per unit to the cost of production to compute sale price. The cost of producing a product is, of course, not a fixed number but depends on the number of units produced. As production increases, fixed costs are spread over more units of production, reducing the cost per unit. In addition to just covering costs, the sale price includes in this case a predetermined profit level or a return on investment in the business.
- *Target return pricing.* The target return approach involves computing a target rate of return on your investment in the business and adding this to the sale price.
- *Perceived buyer value pricing.* With this approach, price is based on the perceived value of the product in the eyes of the buyer. To use this method, the market is segmented as explained in the previous Chapter to identify the type of buyer who will value the product the most. The market segment that values the product the most is known as the target market on which you want to focus your marketing and promotional efforts. Therefore it is important to identify the target market for the product as well as the units of product that can be sold due to changes in the selling price.
- *Type of buyer pricing.* Pricing decisions also depend on whether you are selling to the ultimate consumer of the product or you are selling an ingredient to a processor or manufacturer of the consumer product. Intermediaries such as manufacturers are normally more sophisticated buyers. They are looking for specific attributes and know what they are willing to pay for them. In addition, competitor's prices are more important in the purchasing decision of intermediaries because of the volume involved. In these cases the pricing decision will be more dependent on the price levels of competitors. As a general rule, consumers are less sophisticated in their purchasing decisions and more likely to respond to the emotions of the moment. They are less sure of the value of product attributes and are easier to influence as to their value. Therfore pricing decisions, along with the promotion program, will differ depending on whether you are selling to an intermediary or the final consumer.
- *Price of competitor's products pricing.* The importance of the price of competitor's products depends on whether you are producing a commodity or a differentiated Product. Commodities where every unit of production is the same are highly influenced by the competitors' prices

because buyers see no difference between your product and those of competitors. However, the degree to which you convince buyers that your product is unique or different from those of competitors will influence the degree to which competitor's prices will impact your sales. Differentiation can be expressed in the attributes of the product itself or the services provided with the product;

- *Value pricing.* Buyers make an evaluation of your product by comparing it to competitors' products. This evaluation is based on quality and price. In this instance, quality is defined as how closely the product meets the needs of the buyer. It is important to note that quality doesn't need to be real, it can be imagined. The only thing that counts is that the buyer believes it is different. Also note that price is also an important factor in a buyer's decision. Therefore buyers will prefer a low priced product to a high priced product. So buyers informally take both of these factors into account when evaluating a buying decision by determining a product's value. Value is computed by dividing the product's quality by its price. Because quality is a subjective assessment, companies try to influence the buyers perception of quality. The buyer's assessment of quality is only relevant at the time of purchase.

  As a consequence of the previous paragraph, note that the implicit assumption that high prices and high market share are incompatible is simply incorrect. In a variety of industries, from software to pharmaceuticals, specialised chemicals to cars, aircraft to apparel, it is quite common for the premium price brand also to be a market share leader. High market share and high prices can be achieved if prices truly reflect high customer value [1].

Pricing options related to new products offer additional advantages and challenges. Educating the buyers on what your product is and why they want to purchase it is important. However, pricing your product when the buyer is just learning about it and before you have competitors is unique. There are two typical strategies that you may want to employ:

- *Skimming.* If you are bringing a new product to a target market, a price skimming strategy may be employed [3]. With this strategy you set your price high with the intention of selling to a relatively small portion of your target market – just those high end users who are willing to pay a premium price for your product. Although you don't sell a large quantity of product, your profit margin on each unit is large. A danger of using the skimming strategy is that competitors will enter the market and undercut your price. A skimming strategy works best where your buyers are relatively insensitive to the price level. In other words, demand is inelastic. It also works best in situations where fixed costs are relatively small because fixed costs are only spread over a small number of units. Skimming is sometimes used during the business start-up phase where only a small quantity of the product is produced. As production is ramped-up, the price can be lowered to expand the number of buyers.

- *Penetration Pricing.* Penetration pricing is the opposite of skimming [3]. Penetration pricing involves setting your price low so you can penetrate your target market with a large number of sales and garner a large market share. Once market share is captured, price may be increased. An advantage of penetration pricing is that it will make the industry less attractive to competitors. Penetration pricing works well when the buyer is sensitive to price. In other words, demand is elastic. It is also a good strategy if your cost of production involves high fixed costs because these costs can be spread over many units creating economies of scale. Although this strategy might appear to work for small, value-added enterprises, few will have the infrastructure and size to operate at economies of scale.

Another issue to consider in pricing strategies is the markup pricing by intermediaries. The price you set for your product may not be the price paid by the consumer; the consumer's price may be much higher. If you plan to use intermediaries such as distributors, wholesalers and retailers to distribute and market your product, they will mark up the price to cover their costs.

Finally, it is important to realise that pricing as a dynamic process [4] changes in environmental conditions, in marketing strategy, and in customer needs can require changing selected elements of the pricing process, which in turn can lead to a modification of the prices adopted. Also, and although the objective of the pricing process is to determine a pricing strategy, which will be a basis for profitable decisions in the medium and long term, pricing strategies are always context-specific and thus bound to change. Even global companies, such as DuPont, rarely adopt a truly global pricing strategy, as the specific elements of profitable pricing decisions depend upon local market conditions and country-specific marketing objectives. A profitable pricing strategy in one country might be a marketing blunder in another country [1].

5.2.2 The Pricing Setting Process and Framework

The strategic decision making process involving pricing was explained by Ohmae [5] as a recursive process involving:

- the definition of pricing objectives;
- the analysis of key elements in pricing decisions;
- the selection of profitable price ranges; and
- the implementation of price change.

All pricing decisions should take into account the framework developed in Figure 5.1.

The objectives of the pricing process are a direct result of a company's overall strategy. For instance, a company may pursue a growth strategy of rapidly increasing market penetration and market share. This will require, at least in the short term, the adoption of a different pricing strategy than the pursuit of a strategy aimed at increasing profits over time.

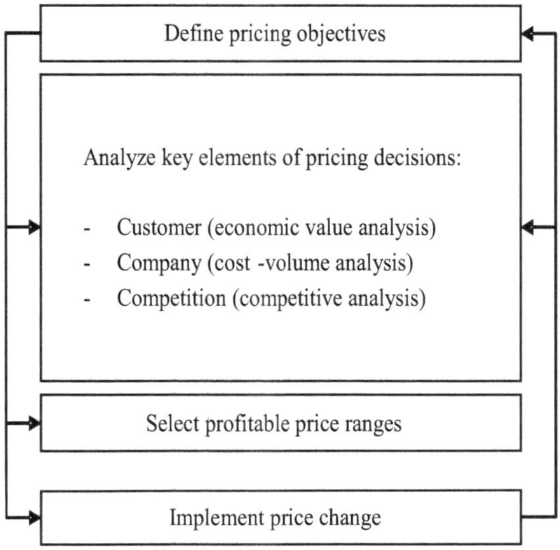

Figure 5.1. Strategic pricing analysis framework (adapted from Ohmae, [5])

Ohmae explained the analysis of key elements in pricing decisions using a triangle. Each of the three corners of the triangle – company, customers, and competition – needs to be analysed and requires the use of specific tools in order to guide profitable pricing decisions [5]:

- Cost Volume Profit (CVP) analysis should be used to capture the company-internal perspective and to understand the implications of price and volume changes on company profitability;
- Competitive analysis to gain insight on trends in competitive strategies; and
- Economic value analysis to understand sources of value for customers.

Economic value analysis is a tool designed to comprehend and to quantify the sources of value of a given product for a group of potential customers. It is clear that it is not always possible to set the price only as a function of the value of a product; however, without knowing a product's value, profitable pricing decisions cannot be made.

The concept of economic (or customer) value is understood in this book – as in Simpson *et al.* [6], or Walter *et al.* [7] – as the difference between perceived customer benefits and sacrifices. Models presented later will model customer value in a dynamic way, considering competition and for different market scenarios.

The process to model customer value can be performed in different ways as will be presented in the next chapters. If we follow the process properly, perceived customer value for our product and its competitors can be calculated. This will later be linked to estimate market share and sales. This customer value or economic value process of analysis will be always a variation of that presented by Hinteruber [1] divided into six steps, as follows:

1. identify the cost of the competitive product and process that consumer views as best alternative;
2. segment the market to see the way in which customers use and value the product and from how they value their respective reference products;
3. identify all factors that differentiate the product from the competitive product and process;
4. determine the value to the customer of these differentiating factors;
5. use the reference value and the differentiation value to determine the total economic value (the sum of these values is used in [Hinteruber];
6. use the value pool to estimate future sales at specific price points.

The process of customer value modelling is a critical process to explore revenue patterns of the enterprise over time and for a given strategy. Revenue dynamics and other factors, as will be explained later, derive the enterprise allowable investments to improve customer perceived value of the product, and therefore, new possibilities to meet enterprise targets in the mid-long run.

5.3 The Cost Structure and the Value Chain

The concept of the value chain was first described and popularized by Michael Porter in his 1985 best-seller, *Competitive Advantage: Creating and Sustaining Superior Performance* (see Figure 5.2). The value chain categorises the generic value-adding activities of an organisation. Once these activities are identified, the costs and value drivers for each value activity can be analysed with the ultimate goal of maximising value creation while minimising costs.

By subdividing an organisation into its key processes or functions, Porter was able to link classical accounting to strategic capabilities by using value as a core concept, *i.e.* the ways a firm can best position itself against its competitors given its relative cost structure, how the composition of the value chain allows the firm to compete on price, or how this composition allows the firm to differentiate its products to specific customer segments.

This value chain analysis and concept has been extended beyond individual organisations and can be applied to entire supply chains of an industry. Capturing the value generated along the chain is the new approach taken by many management strategists to develop new business models, or to create improvements in its value system in other ways. In this manner, we can generate a certain industry value chain representation, where all the actors in a given product supply chain can be considered (see an example in Figure 5.3, taken from a case study that will be presented later in the book).

A recent evolution of the Porter model for supply chains is the SCOR (Supply Chain Operations Reference) model. This model is actually a process reference framework that has been developed and endorsed by the Supply Chain Council (a global trade consortium in operation with over 700 member companies, governmental, academic, consulting groups, etc.) as the cross-industry standard diagnostic tool for supply chain management. SCOR enables users to address, improve and communicate supply chain management practices within and between all interested parties.

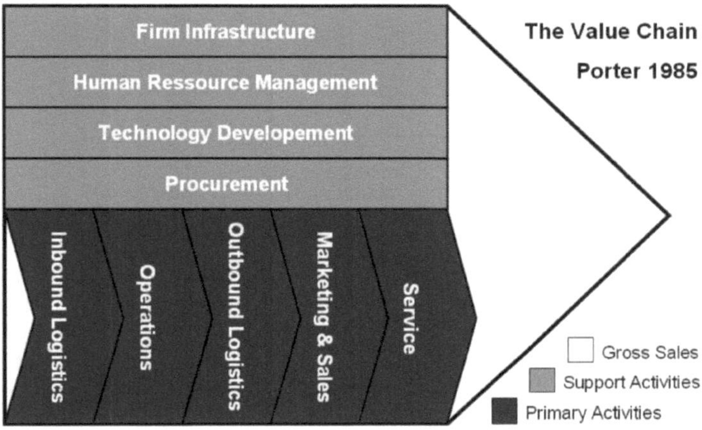

Figure 5.2. Porter's value chain model in 1985

SCOR is a management tool (the reader is referred to [12] for further SCOR model references [8]), a process reference model for supply chain management, spanning from the supplier's supplier to the customer's customer (see Figure 5.4). The SCOR model has been developed to describe the business activities associated with all phases of satisfying a customer's demand. By describing supply chains using process building blocks, the model can be used to describe supply chains that are very simple or very complex using a common set of definitions. As a result, disparate industries can be linked to describe the depth and breadth of virtually any supply chain. The model has successfully been able to describe and provide a basis for supply chain improvement for global projects as well as site-specific projects.

In addition to process elements, SCOR reference frameworks also maintain a vast database of standard process metrics aligned to the Porter model, as well as a large and constantly researched database of prescriptive universal best practices for process execution.

The "SCOR" framework has been adopted by hundreds of companies as well as national entities as a standard for business excellence.

From a strategic *cost structure management perspective*, value chain analysis (and consequently the SCOR model) has three key characteristic attributes that make this technique very useful in analysing current supply chains [9]. These attributes are:

1. a clear identification of the strategy chosen by the organisation;
2. its emphasis on sources of sustainable competitive advantage; and
3. its focus on the importance of complex linkages and interrelationships.

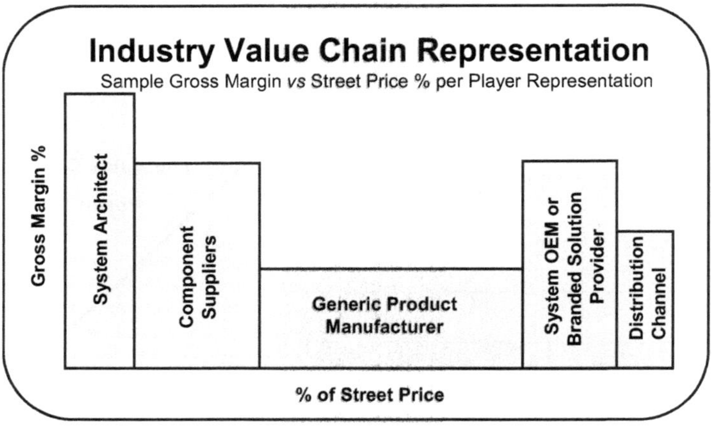

Figure 5.3. Actual sample representación of an industry extended value chain

The first attribute emphasises the need to manage costs of activities and processes in the context of the strategy selected by the organisation, i.e. there should be a strong link between the value chain activities and the organisation's strategy. This might appear obvious but the reality is that many organisations suffer from inertia and, as a result, processes, activities and systems that were deployed a long time ago continue to be performed even though some of them are irrelevant and thus should be eliminated or at the minimum significantly modified. Thus value chain analysis requires an organisation to determine activities that are denoted as strategic value chain activities. These are activities that give the organisation the potential to create value for the consumer as a way of creating and sustaining competitive advantage. Once strategic activities have been identified, the remaining activities must also be clearly identified and denoted as non-strategic. While such activities are important for the overall provision of products and services, they do not confer competitive advantage to the organisation. All non-strategic value chain activities should be streamlined, reduced or outsourced so as to make sure that the organisation's efforts are geared towards activities that have the greatest impact on its ability to achieve and sustain competitive advantage.

The second characteristic of value chain analysis is concerned with what the organisation is good at. Once strategic and non-strategic activities have been clearly understood, an organisation should then identify specific strategic value chain activities that it is good at. The idea here is that an organisation should exploit such characteristics as a way of creating value for the consumer. Even though most activities are necessary in the provision of products and services, not all of them are critical in distinguishing an organisation from its competitors. In order to achieve competitive advantage, an organisation must perform activities in which it has a unique advantage. Once these activities have been identified, creating and sustaining competitive advantage involves closely managing them by making sure that they are not only performed well, but that they are also well resourced. The problem with many organisations is that strategic value chain activities are not identified. To make matters worse, when it comes to reducing costs, an across-the-board approach, *i.e.* cost cutting, is applied.

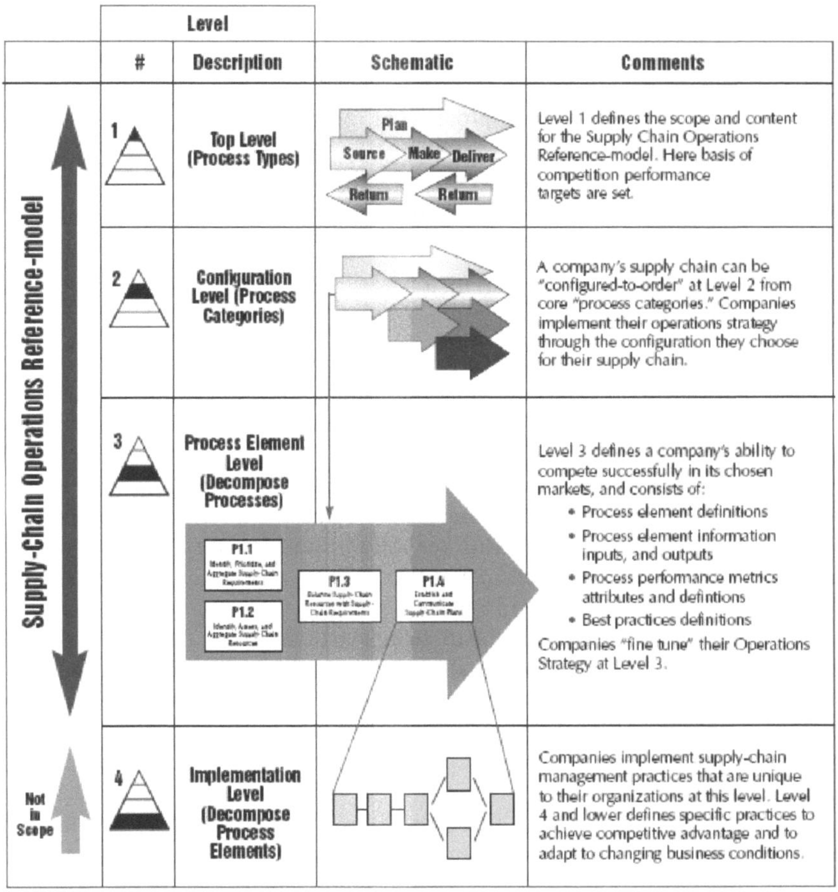

Figure 5.4. SCOR model [8] (SCOR is a registered trademark of the Supply Chain Council in the United States and Europe)

The third attribute of value chain analysis is the emphasis on both internal and external linkages and interrelationships. Internal linkages are typified by relationships between tasks and activities that form a process within the organisation's value chain. Using value chain analysis, an organisation is able to understand the impact of one activity on the performance and costs of another, irrespective of where in the organisation activities are performed. Taking a process view of the organisation facilitates the understanding of an end-to-end (from product design to post-sales activities) value chain activities and costs.

5.4 The Value-driven Planning Process. A Case Study

In the previous chapter – Sections 4.2 and 4.3 – the market segmentation process and the purchasing process were reviewed and analysed for a product of a virtual enterprise named "FPD Co." that was trying hard to go to market with a new product, more precisely, a new flat panel display. Certain segments were targeted and a plan was developed to reach clearly those customers by understanding the purchasing process of potential buyers (where targeted segments and customers work).

In this new case study we review issues related to the "FPD Co." new product introduction, its value proposition and value chain. A critical aspect of the case study is the consideration of strategies to:

1. get to market quickly; and then
2. get the cost curve down.

As the reader may guess, the idea of the organisation through this analysis is to limit the short-term exposure while positioning itself to capture the maximum long-term upside. With this purpose, the organisation tried to establish a plan to capture value over the product rollout time.

"FPD Co." recognised that customer priorities, technologies and business required designs were migrating and therefore new opportunities appeared for value capturing and product differentiation. Besides this, the organisation involved in the case study considered that, adding this image display, products could improve global "FPD Co." business strategy.

At this point the team started to search and to assess new methodologies and/or processes to capture value during the product rollouts. The strategy to capture value was to be presented somehow defining a set of *strategic control points* (as introduced by Slywotzky and Morrison [10]) as well as an overall business model where the value capture points were, as precisely as possible, located.

Strategic control points are those activities that are especially important for achieving strategic objectives. When organisations do not have multiple control systems that focus on strategic control points, they can often experience difficulties that cause managers to re-evaluate their control processes. The purpose of strategic control points is to protect the profit stream that the business design creates from the corrosive effects of competition and customer power. In each industry there are different types of strategic control points that can be identified within a hierarchy.

In this case study, and as a general rule to establish strategy and control points properly, the team recognised the need to:

1. seriously apply market analysis discipline;
2. plan operations specifications;
3. document major assumptions;
4. document key milestones;
5. be careful and limit short-term exposure.

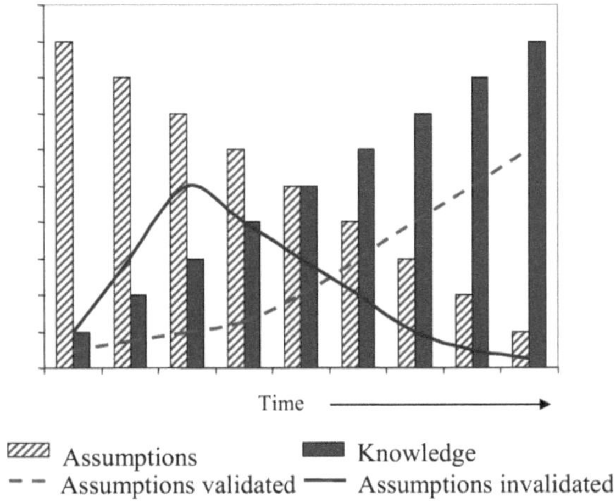

Figure 5.5. A graphical representation of the assumption-to-knowledge ratio

Note that at this stage of strategy development it is extremely important to know how quickly assumptions can be converted to knowledge, and what to do when we invalidate any assumption (see assumption to knowledge ratio graph in Figure 5.5). This was explained by McGrath and MacMillan in their work entitled *Discovery-Driven Planning* published in the Harvard Business Review [11]. Basically, McGrath and MacMillan argue that when weighing a major strategic venture (like a new market or an innovative product introduction) the use of conventional planning tools to manage these ventures may result in very high risk. In these cases, the use of a disciplined process to uncover, test, and (if necessary) revise the assumptions behind venture's plan systematically is required. By doing so, there is exposure to the make-or-break uncertainties common to ventures, and uncertainties can be addressed at the lowest possible cost. McGrath and MacMillan suggest a five-step process for successful venture planning [11]:

1. Bake profitability into the venture's plan. Instead of estimating the venture's revenues and then assuming profits will come, create a "reverse income statement" for the project: determine the profit required to make the venture worthwhile. Then calculate the revenues needed to deliver that profit.
2. Calculate allowable costs. Lay out all the activities required to produce, sell, service and deliver the new product or service to customers. Together, these activities comprise the venture's allowable costs. Ask "If we subtract allowable costs from required revenues, will the venture deliver significant returns?" If not, it may not be worth the risk.
3. Identify assumptions. If there is still the belief that the venture is worth the risk, work with other managers on the venture team to list all the assumptions behind expected profit, revenue and allowable costs calculations. Use disagreement over assumptions to trigger discussion, and be open to adjusting the list.

4. Determine whether the venture still makes sense. Check assumptions against reverse income statement for the venture. If you can still make the required profit, given your latest estimates of revenues and allowable costs, the venture should go forward.
5. Test assumptions at milestones. Use milestone events to test — and, if necessary, further update — assumptions. Postpone major commitments of resources until evidence from a previous milestone signals that taking the next step is justified.

For instance, continuing with the case study, in "FPD Co." basic assumption for the product was that a $200 product cost (at volume) was possible for the first product. With this cost, a $450 street price and around 30,000 units/month, business could break even 1 year ahead of regular enterprise established 3-year boundary condition.

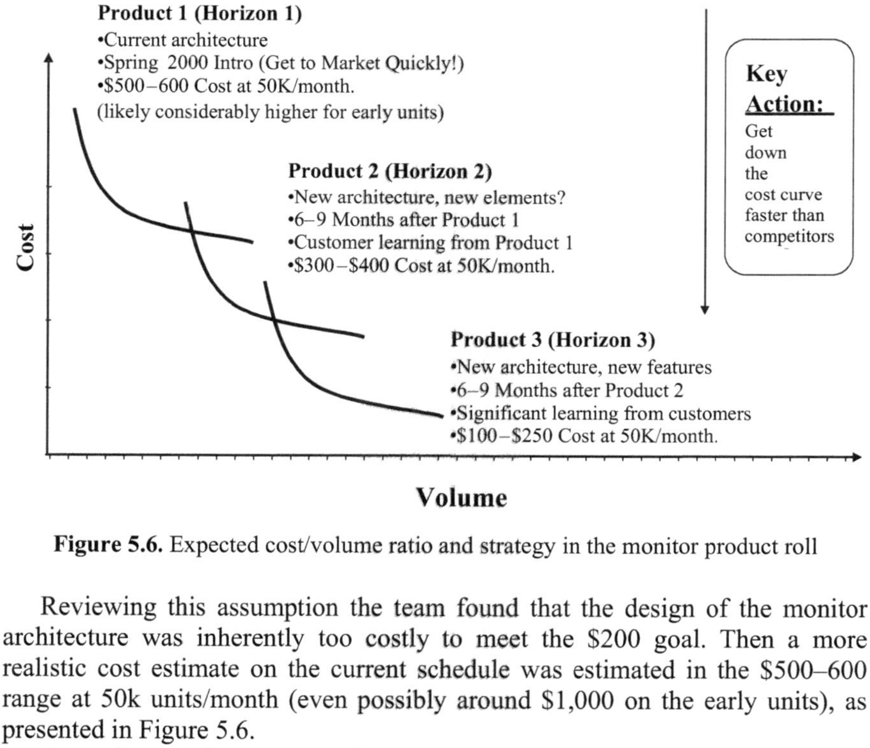

Figure 5.6. Expected cost/volume ratio and strategy in the monitor product roll

Reviewing this assumption the team found that the design of the monitor architecture was inherently too costly to meet the $200 goal. Then a more realistic cost estimate on the current schedule was estimated in the $500–600 range at 50k units/month (even possibly around $1,000 on the early units), as presented in Figure 5.6.

Once the previous assumption was revised and updated, the team also considered new assumptions. The top candidates were as follows:

- *New assumption candidate #1*: A target market which will sufficiently value the features of the first product would pay enough to cover the product cost (leaving a modest per unit Gross Margin). A design team was to be launched concurrently to drive the cost down for a 6–9 month product roll.

- *New assumption candidate #2*: No target market will pay enough for the first product to cover the product cost, but the value of getting a product quickly to market to begin learning is sufficiently valuable to go to market with minimal per unit losses. In order to keep loss exposure manageable, a limited supply of product #1 would be built (enough to learn and keep the supply base engaged). Also a design team was to be launched concurrently to drive the cost down for a 6–9 month product roll.

In a meeting, strategies to limit the short-term exposure, and position to capture maximum long-term upside were discussed and the following actions to be examined classified as follows (see expected profit over time and product roll in Figure 5.7):

- Strategies to limit the short-term exposure:
  - partner to limit development costs;
  - share the Start-up downside by developing a way for partners to also share in the later upside;
  - learn from partners to minimise risk;
  - for the High-Cost Product #1, either find a target market which sufficiently values the features to pay above the product cost, or limit volumes if the initial gross margin is negative;
  - develop metrics to measure progress toward achieving the objectives under the strategies of limiting the short-term exposure and positioning to capture maximum long-term upside;
  - quickly roll to Lower-Cost Product #2.
- Strategies to capture maximum long-term upside:
  - determine and establish strategic control points;
  - hit program milestones to install confidence in key stakeholders;
  - get to market quickly to:
    - develop customer relationships;
    - begin learning and establish market momentum;
  - design to a product cost that will enable flexibility to price low and still be profitable over the long run;
  - develop effective and efficient delivery system;
  - focus on positioning for significant EVA at the appropriate stages. Not measurement of EVA during the start-up stages;
  - keep supplier base focused on our business.

Pre-evaluation of strategic control points was presented as a departing point of this analysis, the work was done adapting those in Slywotzky and Morrison (see page 53 in [10]), adding a horizon and key actions to ensure said control. The result was the table presented in Figure 5.8, where several potential strategic control points are listed according to the scheme presented by Slywotzky & Morrison. They mention that every good business design has at least one strategic control point, and that they have found that the best business designs have two or more. In this case study, the team decided to try to establish several if possible. The table in Figure 5.8 includes potential control point for the business of the case study in the shaded rows.

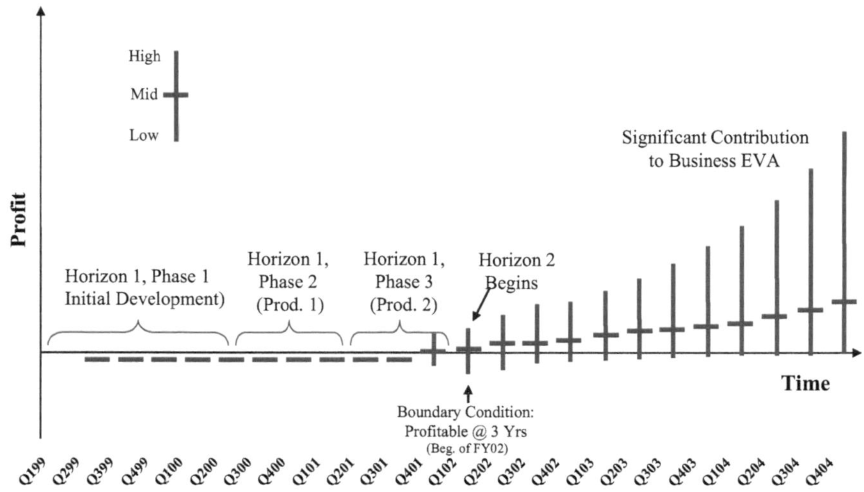

Figure 5.7. Expected profit over time and product roll

To conclude the case study, it was interesting to see how and where these strategic control points could fit within the business model. The team discussed the possibility of capturing value points at two different positions of the value chain. These two value capture points would include specific strategic control points as follows:

- *Value capture point #1: l*ocated at the system architect. Includes the following strategic control points:
  - "Own the Standard";
  - "Manage the Value Chain";
  - "Own the Patent"; and
  - "Product Development Lead".
- *Value capture point #2: l*ocated at System OEM or branded solution provider and at the distribution channel. Includes the following strategic control points:
  - "Manage the Value Chain";
  - "Own the Customer Relationship"; and
  - "Own the Brand" Strategic Control Points.

This first definition of the value capture points and of the strategic control points for the business model is presented in Figure 5.9. In that figure we use the structure of the graph in Figure 5.3, which was originally designed for this purpose and business model.

Profit Protection	Strategy	Example	Potential Business Control Point	Timeframe	Key Actions
10	Own the standard	Oracle, Microsoft	Own display standard Own superior system architecture	Horizon 3-4	Establish the "display company" to develop superior product configuration
9	Manage the value chain	Intel, Coke	System architecture Global information backbone Financial transaction backbone	Horizon 1-4	Develop superior architecture Develop information and financial supply chain backbones
8	Own a string of super dominant positions	Coke Int'l McDonald's Int'l			
7	Own the customer relationship	GE, EDS	Good understanding of what customers value Consistent delivery of valuable solutions	Horizon 1-4	Develop customer relationship Develop compelling solutions
6	Own the patent, brand or copyright	Mattel, Sony	Imaging brand	Horizon 3-4	Secure patent Develop brand
5	Posses a significant product development lead	Intel			
4	Posses some product development lead	Sony	First product to volume First to future enhancement	Horizon 1-2 Horizon 3-4?	Get to market soon Continue to innovate
3	Sell a commodity with a 10%-20% cost advantage	Nucor, Southwest Air			
2	Sell a commodity with cost parity	Countless			
1	Sell a commodity with a cost disadvantage	Countless			

Figure 5.8. Suggested strategic control points (adapted from [10])

To conclude this business case, we have to say that the strategy development team decided the following actions to continue the case:

- learn more about the display industry;
- complete System Dynamics and Supply Chain modelling;
- continue Discovery-Driven Planning;
- develop short-term divisional metrics;
- perform quantitative analysis, including analysis of financial implications of different price, cost, feature, and schedule combinations;
- refine strategies for strategic control and value capture.

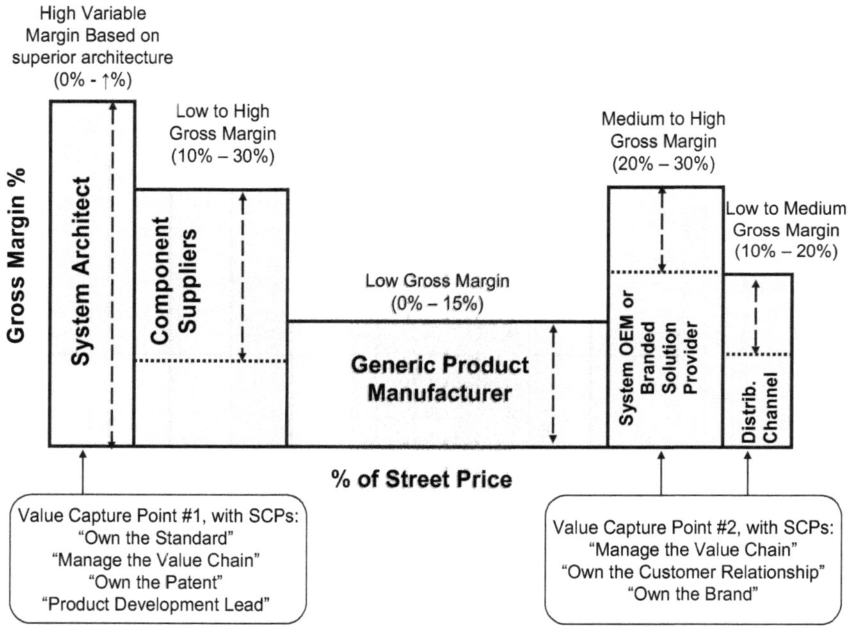

Figure 5.9. Business model, value capture points and strategic control points

5.5 References

[1] Hinteruber A, (2004) Towards value-based pricing – An integrative framework for decision making. Industrial Marketing Management, 33: 765–778.
[2] Malhorta N, (1996) The impact of the academy of marketing science on marketing scholarship – An analysis of the research published in JAMS. Journal of the Academy of Marketing Science, 24(4): 291–298.
[3] Lamb C, Hair J, McDaniel C, (2000). Marketing. (5th ed.). Cincinnati, OH: South-Western College Publication.
[4] Shipley D, Jobber D, (2001) Integrative pricing via the pricing wheel. Industrial Marketing Management, 30: 301–314.
[5] Ohmae K, (1982) The mind of the strategist – The art of Japanese business. New York: McGraw-Hill.
[6] Simpson P, Siguaw J, Baker T, (2001) A model of value creation – Supplier behaviours and their impact on reseller-perceived value. Industrial Marketing Management, 30:119– 134.
[7] Walter A, Ritter T, Gemuenden HG, (2001) Value creation in buyer – seller relationships – Theoretical considerations and empirical results from a supplier's perspective. Industrial Marketing Management, 30: 365–377.
[8] Supply-Chain Council (2008) Supply Chain Operations Reference Model (SCOR Model). Version 9.0. http://www.supply-chain.org.
[9] Chivaka R, (2007) Strategic cost management: Value chain analysis approach. Accountancy SA, August.
[10] Slywotzky J, Morrison DJ, (1997) The Profit Zone. New York: Times Business.
[11] McGrath RG, MacMillan IC, (1995) Discovery-Driven Planning. Harvard Business Review. 73(4): 44–52.
[12] http://www.supply-chain.org

6

Understanding Hi-tech Business Growth

6.1 Characterising Hi-tech Business Planning Process

In the hi-tech business environment, the business planning process will try to reach certain objectives considering the needs and wants of the customers, shareholders, and other stakeholders. These general business objectives can be placed into four groups: profitability, growth, risk and social objectives:

- Profitability is, as a general rule, a priority. It is the necessary condition that allows us, in the long run, to reach the other objectives.
- Growth can be important at different moments of the product life cycle, for instance, in high-growth markets gaining share is easier and more valuable, it reduces pressure on price, it ensures access to technology, it deters subsequent entrants in the market, *etc.*
- People, environment and safety.
- Many companies claim that they have social objectives to fulfil. They actively want to contribute to the discussion of socially relevant issues by engaging in dialogue with interested sections of society.

Achieving these business objectives requires a business strategy. The strategy setting process may follow standard organisational planning methods, which normally include (see Figure 6.1):

- Deriving from corporate goals the business objectives and policies.
- Determination of current business performance.
- Determination of the target performance measures (Key Performance Indicators – KPIs). Improvements will be made based on accepted business, user and SCM performance indicators.
- Establishing principles to guide strategy implementation by means of planning, execution, assessment, analysis and improvement.

In hi-tech businesses, relational input is important where projections of both market demand and competitive position are essential inputs to strategy [1,2]. There is simply not a large enough sample of good data to get statistically valid outcomes on the basis of projections from past trends and patterns, nor are there controlled representative data sources to support correlations or regression analysis.

Figure 6.1. Business strategy model

As a general rule, business planning within a hi-tech environment is both dynamic and complex, with a critical need for nonlinear, relational input and mathematical rigor. This is particularly the case where planners and decision-makers must rely on subjective and potentially biased data [3], and where data sources span across cultures and languages.

For all these reasons, planners increasingly turn to simulations to build confidence and consensus in selecting operational investments to improve or protect metrics such as market share, revenue and profit for global hi-tech businesses. Adding the ability to analyse decisions in light of the impact on share, revenue and profits turns the simulation model into a decision support system.

The reader has to understand that many hi-tech planners are more interested in share as a business metric than either revenue or profit. This is closely tied to the fast pace of technology and product life cycles, and the increasing difficulty of trying to gain market share as the market matures. In addition, market share is tracked and reported in trade and investment publications and watched closely by investors and analysts looking for visible short-term results to publicised strategy. However, note that although market share may be widely used, it can sometimes be a very poor performance metric. Absolute sales volume could be preferable, since it is directly traceable to customer gains and losses. For instance, 90% of a tiny market could contribute less to earnings than 25% of a large market

As a first step in the introduction of different models that can meet these needs, we shall summarise the unique characteristics of the hi-tech marketplace:

- volatile, uncertain markets with great pressure on managers for near-term market share and/or financial performance (in the U.S. hi-tech programs and product lines may be funded for a period of time in spite of poor financial results if they prove themselves, quarter by quarter, able to capture and hold share in strategic markets).
- multiple planning dimensions, including technology path, product architecture, delivery chain, alliances, channels, and services;
- little historical data, due to technology adoption rates, reorganisations, mergers and acquisitions, globalisation, and new channels for order and distribution.
- isolated groups of expert knowledge, each with their own language and systems.
- absence of a single view of the possible impact of an investment, especially when results are scattered across space and time, well beyond the scope of any single enterprise planning system.

6.2 Hi-tech Business Growth. A Case Study

In the following paragraphs we describe all the vicissitudes experienced by a model development team, using the System Dynamics approach to integrate business disseminated structural concepts into a current powerful simulation tool for business planning.

Although the final outcomes of the project – how business conducts its business planning today – were very well ranked within the organisation, it was probably the insights and lessons gained during the process, and about the process, that were the most outstanding assets for those who participated in the project.

6.2.1 Reasons for this Modelling Effort

This case study took place in a business – we will call it "Print Co." – of a well known hi-tech corporation. The business offered a wide variety of printing supplies to customers, who were mainly LaserJet printer owners. In 1997 a dynamic modelling development team was created and since then has joined different business projects. The common overall objective of these projects was to gain understanding of the business using a systemic approach. The results would be measured in terms of clear improvements of the business planning process through a more holistic view of the business and a much better understanding of the investments implications.

Business planning within a hi-tech environment is a very complex process. Several reasons can be found for this:

- Many potential business scenarios. This makes business results forecasting more difficult, increases business risk, and forces business teams to reduce the percentage deviation between actual and projected results.
- Many different aspects to be taken into account: financial constraints and investments alternatives, current customer profile definition and importance, market place and competition status, product and process technology fast development, current relationship and partnership with suppliers, economy globalisation implications, ...etc.
- For many of these aspects there is a lack of previous historical data. Some examples are: Internet development impact on printing (customers using printers for different purposes than previously and at different usage rates), new emerging technologies for printing (cheaper and easier printing), emerging sectors or markets where printing technology may impact importantly (editorial companies, bookstores, photo machines and pictures, etc.), emerging technologies substituting printing (laptops reducing the amount of pages that you print since you bring the laptop with you), etc.
- Many fields of expertise required to fulfil the information requirements, and therefore many different teams dedicated to different research topics.
- Combining all the information, putting things together, is a hard job. This is normally a consequence of the process design by itself. Teams involved in the process focus their attention to particular aspects of the problem, this specialisation may sometimes lead to a lack of general understanding of the global business process, and therefore, decisions involving policies driving to solve local, instead of global problems, are likely to appear. As an artificial example, we could have a team developing alternative policies to increase the usage rate of printer toner without considering the sensitivity to price or to environmental issues that the customers might show if the usage increases.

6.2.2 Fuzzy and Soft Marketing

Some years ago, "Print Co." had released three different products (toner cartridges) with extremely high expectations:

- The first one was an "EX cartridge" that was announced as a cartridge providing more yield than a normal one, and also a fraction more expensive.
- The second product was a "low yield cartridge" marketed as cheaper than the normal cartridge; the third one was a "Multipack" including several units that the customer could pick at a reduced price. Internal business expectations did not match customer ones. Although customers seemed to be product price sensitive, the cartridge yield – number of pages that

can be printed per cartridge – did not turn out to be a relevant aspect to condition their purchasing decisions, at the same time, purchasing more than one cartridges created storage problems for the customers.

In summary, these programs never did match the "Print Co." expectations; the fact that marketing decisions were based importantly on fuzzy or soft customer knowledge was a main concern.

"Print Co." considered that moment as an inflection point. The organisation had to learn quickly from these past experiences, and the business had hardly worked on the process of "building more intelligence" to support better marketing decisions.

This case study illustrates the process that the organisation went through, focussed on the better understanding of the particular preferences of the existing and potential business customers, and how those customers perceived the value of the products and service attributes differently.

6.2.3 Understanding the Business Process Better

It was at this time, when the "Print Co." Marketing Manager met several analysts from the Corporate Consulting and System Modelling group (CCSM group) for the first time. CCSM group was part of the corporate organisation, and its main approach to support products, process and relationships was systems dynamics modelling. CCSM was brought in, and the idea was to conduct a workshop to put together a report of what the "Print Co." business process looked like. CCSM worked for a period of 6 months with several sub-teams. A different part of the business process was analysed in depth with each sub-team and at the same time and through CCSM, a high level vision of the overall business process was diffused to all of the sub-teams.

By understanding each part of the business, capturing and diffusing what was learnt to the rest of the sub-teams, the construction of the "Print Co." business process structure was possible. Subsequently, the very first "Print Co." business model version was born.

The main value of this first model was that everyone in the different teams validated it, consensus was reached, and therefore people within "Print Co." could be educated using the model. This first version was known as the "spaghetti diagram of the business"; the reason for this was the shape of the diagram, which contained an important number of arrows linking the names of the different variables with cause and effect implications. Although the model was still far from formalised, the emerging opportunity to "tell the Print Co. business story" using the model, was found to be absolutely brilliant. The overall process, involving the whole organisation, was easily transmitted to employees. The implications of local decisions for the overall business performance could be understood and transferred to the people with far more ease. A much more dynamic understanding of the business was now enabling people to capture the link between structure and business behaviour.

Early in January 1998, the "Print Co." business model was accepted as a suitable representation of the business process structure; previously the model had gone through a structural validation process around the organisation with remarkable robustness and success.

By that time, the "Print Co." management team was facing important decisions concerning different strategies; some of them were involving issues such as product brands and labels, which were considered as involving high risk for the business. The General Manager had said that he wouldn't support those strategies unless substantial analysis around them had been done; he remarked that "Print Co." couldn't even support them internally, and would have to get some business external help (from other businesses within the corporate organisation) to support them. All this would be required before "Print Co." could go to suppliers to work out the strategy choices.

The management realised that building quantitative analysis capabilities within the organisation was an urgent need. Qualitative information contained in the "Print Co." business model had been a first and consistent step, but using that work for the formalisation of the model in a powerful computerised simulation tool seems to be the next challenge to face. The results of this work should be not just describing the business process but also explaining and quantifying to a certain extent the business dynamics under possible scenarios.

6.2.4 Understanding the Requirements of a Business Process Model

MH, a "Print Co." business analyst, had joined the model development team some months before, and it was decided that he would be the link between "Print Co." and the CCSM modelling people. In the end, he certainly became being much more than just a 'link'. He was a converted "system dynamicist", a fan and practitioner of this modelling methodology with a more than reasonable knowledge of the software that was being used. His confidence in the work and positive support for several months was critical for the final success of this particular project. On December 15th 1997, MH gave an update on the steps that had been followed to build the simulation model to date. The "Print Co." General Manager was very supportive of the systems modelling effort update, but his major concern was "to do some pilots and test marketing soon, do a controlled experiment – drag out over multiple months". His vision was something similar to a dashboard, "something we can put in our laptop and play with it, something programmable to see how the business is running".

The model development team was moving the work forward extremely fast. Concurrently, a data request was passed on to "Print Co." in order to do the data assessment at the same time as the model formalisation was being completed. The data to be found had to be related to the model exogenous variables, model constants and parameters, but also with any other model variable that they had access to. All possible data were welcomed to validate model simulations later.

The assessment did not turn out to be a trivial issue. It was found to be a very slow and difficult process, depending on many business sub-teams and/or on eventual circumstances within the organisation. Although MH was speeding up the process, the team had to acknowledge that this part of the project would require a higher investment in time and a much more explicit process design than had been initially expected.

6.2.5 Introducing the Marketing Intelligence Team

The team agreed to work using templates, spreadsheets designed specifically for the project, containing the information concerning the model variables to assess. After 2–3 weeks the first templates were returned with some data, but problems continued. Here are some of the causes mentioned:

- No information at all was found for some variables considered essential to produce quantitative results with the model, for instance, usage rate and its drivers.
- Most of the data related to marketing variables was unknown at that time. Although "Print Co." had already decided the segments to compete, there was a clear lack of quantitative information about most of the aspects defining the preferences of the different segments.
- Information about the marketshare in the segments, and other core customer measurements, was fuzzy, and there were no more than "educated guesstimations" about metrics offering information regarding current and past "Print Co." value proposition to the different customers.
- Some financial data was not organized as desired for modelling purposes for instance, linking financial data to product attributes was found to be impossible. At the same time, for example, linking financial data with investments constraints policies within "Print Co.", which had to be formalised in the model, was very time consuming.

By mid-February 1998, the first simulations results were produced and validated. That was certainly exciting and an extraordinary professional experience for the business analysts in the team. However, the team realised that there was a need to focus on gathering market intelligence. This would enable "Print Co." to validate assumptions that had already been made within the current model. This would lead to additional model development once the work had been completed. With this in mind, the market intelligence group was formed within "Print Co.". This group was in the process of outlining research needs, how to synthesise that information, and what would that cost. Towards this objective the system modelling work had already paid dividends in highlighting weaknesses in "Print Co." customer knowledge.

6.2.6 Validating the Model and Preserving the Chain of Belief

When the formalisation of the model equations was completed, and the first simulations results were obtained, the team went through a second validation process (we remind the reader that the first one was mainly focussed on the structural aspects of the model), dedicated to studying how well the model could replicate real available business data.

Once the formulation of the model variables (equations) was a process involving only certain particular members of the team, the rest of the members were really able to appreciate a set of slides mapping the input data used in the model, explaining the algorithms designed for each piece, and showing the output data obtained for each variable compared to real business data. This process was found to be really helpful for the whole development team and has been repeated each time the model has been presented or introduced to a group of people ever since. The team considered that, by doing this, it was preserving "the chain of belief" built along the modelling work.

During this second validation process the team noticed that many of the initial "guesstimations" regarding model unknown variables and parameters were far from the orientative values offered by the model. For instance, cartridges usage rate initial estimations were much lower than what the model was suggesting. In summary, through this inductive process, new weaknesses of current market understanding were found and corrective actions were introduced.

By the summer of 1998, the model was able to predict revenue over time, based on investments in product attributes that the business could afford with the on-going growth reached. At the same time, the share obtained with the model in the different segments was matching existing and new market research data.

At the end of July 1998, the model was used by the management team to analyse the implications of different strategies regarding margin discounts to the resellers. Here are some of the comments from the business general manager (GM) and the financial controller (FC) after the team presentation:

> **(GM):** *Especially now. We have all kinds of big moves – the model can help us as we're brainstorming. It's important for people to see the power of the model – what it's really capable of. We just need to have more discussions like this. This is exactly the kind of discussion we need to have. I wish you could have seen us an hour ago – we just spent a whole hour talking about this – whether margin discounts are working as well as we think. What other questions do we need to ask? What are other scenarios? Who are the people involved? Margin is a great example – it's starting to reach the end of its life. But the important thing to me about the model is that it generated the discussion.*
>
> **(FC):** *What helps me to "get" this model is to feel credibility with it. It's not always clear to me how you "drill down". We always think in linear terms. MH calls this a "chain of belief" – being able to trace the contents of the model, and behaviour, back to what you believe.*

In October 1998, a workshop was conducted at "Print Co."; the idea was to have different teams competing using the model as a tool to evaluate possible marketing strategies. The goal of the teams was to maximise the Economic Value Added (EVA) over a 5-year period using different marketing strategies. Again, the exercise with the model was found to be very powerful and the attendant's feedback was excellent.

Although model results were very useful throughout 1998, at the end of the year "Print Co." people realised that "what they thought the business was, is not what the data was telling them". The need for some model structural changes was acknowledged, and "Print Co." realised that that would be a permanent on-going process in the future.

6.2.7 Concluding Remarks of the Case Study

Where was "Print Co." after the modelling effort, compared to where it was before?

In "Print Co." a system dynamic model was developed to help with the business planning process; by developing it and using it, the organisation could make strategy changes along the line. But when "Print Co." presented its policy to approach business through marketing after that time, something had changed; something was different from a couple of years before. Something definitely related to the way "Print Co." connected marketing investments to market share, the way the "business loop" was closed, the gained intelligence about the business process, the market, and the existing and potential customers.

When this case study was about to be written originally, the business management team was receiving the goals for the results projection over the next few years. These goals seemed to be a very tough challenge for "Print Co.". The model was being used to experiment on whether the goals were reachable, or not, through the current business paradigm. As an example of the previous paragraphs, the organisation found important insights regarding situations constraining the gain of the reinforcing loop that was shown in Figure 3.1. In the current paradigm, revenue growth was allowing a set of marketing investments in the attributes of a family of products, printer supplies, which at the end of the day increase the revenue again. Since this paradigm was currently built into the System Dynamics model, this was being used to try feasible alternatives of investments strategies, with the idea of testing whether projected results could be replicated.

"Print Co." was finding certain market constraints; the model showed that there was "not enough room" in the market to meet results projection within the current business scheme. The need for a redefinition of the business paradigm was appearing.

Transitory processes are a fact in businesses nowadays. Gaining the required market intelligence faster, when change is upon us, is a clear competitive advantage. For "Print Co.", this process took approximately 4 months and $80,000 during the first year. The following year, the team redefined the scenario structure and updated market information, but the cost was only 3 months and $45,000. Business planners estimated that the resulting optimisation could generate annual incremental revenues of $190 million, which is the benefit of a new way to approach business through marketing at "Print Co."

Dynamic modelling transformed knowledge into strategies for increasing customer and shareholder value. It was said that that was the result of a "growing edge" technology – that is, people learn and simulate how the business and market behaves, then apply that framework to a specific situation, which then generated better results, experience and research, which more rapidly enhance the simulation, and so it proved.

6.3 References

[1] Shaw J, Subramaniam C, Tan GW, Welge ME, (2001) Knowledge management and data mining for marketing. Decision Support Systems. 31: 127–137.

[2] Simon H, (1994) Marketing science's pilgrimage to the ivory tower, in: Laurent G, Lilien GL, Pras BD (Eds.), Research traditions in marketing. Boston: Kluwer Academic Publishing.

[3] Senge P, (1991) The fifth discipline: the art and practice of the learning organisation. New York: Double Day.

7

Modelling a Hi-tech Business Growth

7.1 Model Overview

In Chapters 4, 5 and 6 several case studies are presented. These case studies attempt to illustrate typical processes that can be found in "front end" units of hi-tech organisations when these units try to link customer value to business targets. The purpose of these cases was also to show where, when and how some positive experiences using dynamic models to deal with these type of issues took place.

In this chapter we move forward to describe the building and formalisation processes of dynamic models, similar to those which were used in the case studies previously presented.

When working with front end units of hi-tech companies, the general model structure that we presented in Figure 3.1 was then synthesised and refined. The results that we present in this chapter, in terms of confirmation and validation, were discussed with the following business players:

- commercial and consumer business managers;
- systems analysts;
- critical part contract managers;
- financial executives; and
- experts in hi-tech workforce collaboration.

The results are represented by the three sub-models that are shown in Figure 7.1:

- the purchasing behaviour model;
- the financial model;
- the investment model.

In the purchasing behaviour sub-model, "purchasing" represents the customer's decision to buy, and purchasing behaviour refers to the customer response to perception of value relative to the competition.

The question that this sub-model tries to solve is: How does the customer perception of product quality and price attributes impact market share for this business and for its competitors? In the subsequent sections this problem is studied by formalising the relationship among the variables involved.

The financial sub-model deals with how a product and market strategy impact business revenue, and how this revenue is linked over time to the product's price attributes and profit.

The investment model is related to planning and tracking targets throughout the fiscal year by calculating the rate of investment that the business should direct toward a given market opportunity in order to reach its profit goals. The model helps to set up a policy to determine the rate of spending we can accomplish. Also, the model may give guidelines to find out which variables should drive decisions about continuing or changing program investments.

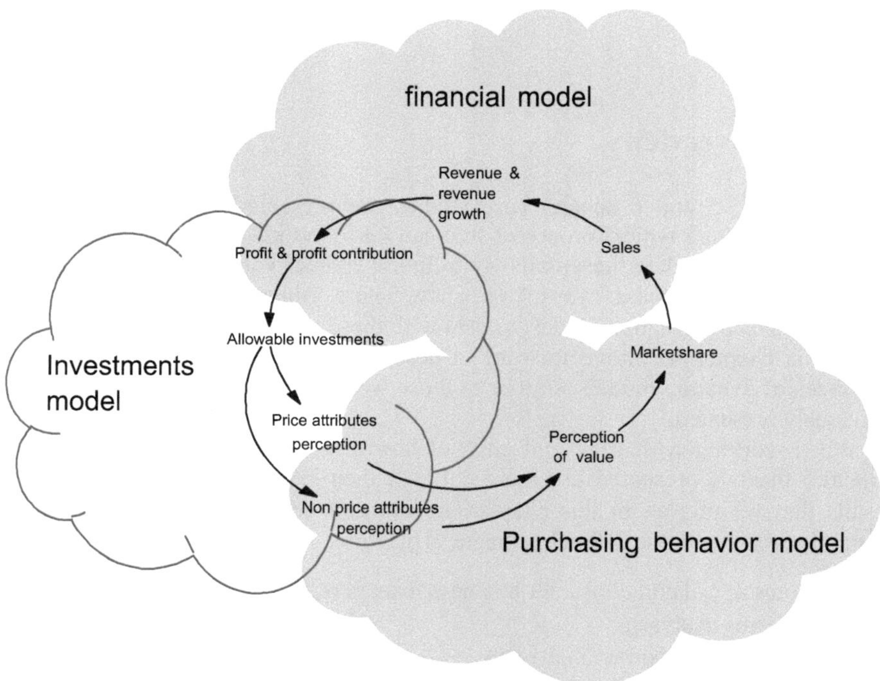

Figure 7.1. Overall business growth model

7.2 Modelling Customer's Decision To Buy

Customers have an open mind with regard to new products and services that come out in different periods of time. The product will have be able to meet their needs and demand and also cater to their preferences.

How does the customers' perception of product quality and price attributes impact market share for this business and for its competitors? We will now attempt to study this problem by formalising the relationship among the variables involved.

Before proceeding with the model development and discussion, we will first describe the notations and definition of the main purchasing behaviour model variables as follows:

Subscripts:

$j= 1,\ldots,N$	Competitors, including this business
$s= 1,\ldots,S$	Market segments grouped by the most important attributes
$i= 1,\ldots,L$	Quality attributes
$k=1,\ldots,M$	Price attributes

Input: Customer perception of each competitor

$Qac_{j\,t}^{i}$	Perceived quality attribute i of the competitor j in t
$Pac_{j\,t}^{k}$	Perceived price attribute k of the competitor j in t
Qab_{t}^{i}	Baseline perception of quality attribute i for all competitors in t
Pab_{t}^{k}	Baseline perception of price attribute k for all competitors in t

Input: Expected impact for each competitor in each segment

$Qa_{j,s\,t}^{\;i}$	Competitor j impact on value for customers of the s segment and through the quality attribute i in t
$Pa_{j\,s\,t}^{\;k}$	Competitor j impact on value for customers of the s segment and through the price attribute k in t

Calculations: Basis for comparison between competitors

$ICP_{j\,t}^{s}$	Index of customer in segment s perception of competitor j

Calculations: Result of investment conditioned by share (reach)

$\Gamma q_{s}^{\;i}$	Elasticity of the quality attribute i for segment s
$\Gamma p_{s}^{\;k}$	Elasticity of price attribute k for segment s
$Pc_{s\,t}^{\;j}$	Presence of competitor j in segment s in t
TCI_{t}^{s}	Total competitor index in segment s in t

Output: Market share change in units of solution product

$MSH_{j\,t}^{\;s}$	Market-share of competitor j in segment s in t
$MST_{j\,t}$	Market-share trend of competitor j in t

7.3 Modelling a Customer Perception of a Product

The model can now be explained as follows. A purchaser (it could be a consumer, but also a technical or procurement manager) will most likely select a product according to widely-held perceptions about its quality ($Qac_{j\,t}^{i}$) and price ($Pac_{j\,t}^{k}$) attributes. Of course we assume that product attributes are characteristics by which products are identified and differentiated. Product attributes usually comprise features, functions, benefits and uses. In our case, examples of quality attributes include:

- reliability;
- ease of purchase;
- scalability;
- network friendliness;
- service availability; and/or
- connectivity.

As the reader may notice, the word quality is used in this section in abroad sense, including product aspects (e.g. reliability) and non-product aspects (e.g. service availability). It can be said that quality encompasses here how well a seller meets all of the nonprice needs that affect the customer buying decision [1].

Examples of price attributes include:

- rebates;
- promotional discounts;
- cost per instance of use; and/or
- channel discounts.

Once the purchaser establishes these preferences for the products of the different competitors, we can define the baseline perceptions as follows:

$$Qab_t^i = MIN_j\,(Qac_{j\,t}^{i}),\ \ with\ j = 1, \ldots, N \tag{7.1}$$

$$Pab_t^k = MIN_j\,(Pac_{j\,t}^{k}),\ \ with\ j = 1, \ldots, N \tag{7.2}$$

Next, we can formalise how much each attribute is able to impact on the value provided by the product to the purchaser, as follows,

$$Qa_{j,s\,t}^{\,i} = (Qac_{j\,t}^{i} \,/\, Qab_t^i)\text{\^{}}\Gamma q_s^i \tag{7.3}$$

$$Pa_{j,s\,t}^{\,k} = (Pac_{j\,t}^{k} \,/\, Pab_t^k)\text{\^{}}\Gamma p_s^k \tag{7.4}$$

In Equations 7.3 and 7.4 we assume that a purchaser in a segment will pay special attention to the attributes of the product most important to that segment. This concept is formalised through an index of elasticity for each price and quality attributes: Γp_s^k and Γq_s^i respectively (each elasticity value is calculated through the model calibration process, and then its value is maintained for the rest of the simulations). Switching costs and other factors may cause customers to be less responsive to changes in some attributes – this is represented in the model as the inherent elasticity of a quality or price attribute in a particular segment.

7.4 Modelling Competition. Value Provided and Perceived

Buyers choose the best preceived customer value – that is the best perceived quality for the perceived price – they can find in the marketplace. Once the impact on the value provided by each attribute of the product is calculated, we can formalise an index that compares the value provided by each competitor's product, as follows:

$$ICP_{j\,t}^{s} = \prod_{i=1}^{L} Qa_{j,s\,t}^{\;i} \prod_{k=1}^{M} Pa_{j,s\,t}^{\;k} \tag{7.5}$$

Assessment of these indexes is not difficult since customer perception of their products is tracked one way or another by most firms [2]. After that, the model simulates behaviour for a given business by showing that the model generates correct changes in individual competitor market share for changes in value (relative to the competition), which can be validated by historical data. It is our main assumption that we can thereafter estimate the share by defined segment for each of the competitors by comparing their customer perception indices, and by assessing their presence in the marketplace *($Pc_{s\,t}^{\;j}$)*, as follows:

$$TCI_{\;t}^{s} = \sum_{j=1}^{N} Pc_{s\,t}^{\;j} \times ICP_{j\,t}^{s} \tag{7.6}$$

$$MSH_{j\;t}^{\;s} = (Pc_{s\,t}^{\;j} \times ICP_{j\,t}^{s}) / TCI_{\;t}^{s} \tag{7.7}$$

Presence of the competitors in the market is related to their reach in each segment. Market reach can vary from very monopolistic to very competitive, or even an almost non-existent reach in any segment.

Equations 7.6 and 7.7 are therefore introduced to model competitor market share in a market where competitive effects are differentially and asymmetrically distributed. Notice how this model can be considered as a simple *attraction* model [3] based on the hypothesis that a competitor market share is equal to its *attraction* relative to all others (Equation 7.7). In our case, competitor's *attraction* in a segment is estimated by ($Pc_{s\,t}^{\;j} \times ICP_{j\,t}^{s}$).

The purchasing behaviour model presented here was designed by modelling teams, as presented in Figure 7.2, where three important competitors (or competitor proxies, where a proxy defines a competitive strategy) were considered..

Share here represents the percentage of target market segment sales that can be expected to flow to each competitor over a given time period, knowing that all the factors are continuously changing and influencing each other during that time. Overall market size remains exogenous to the model.

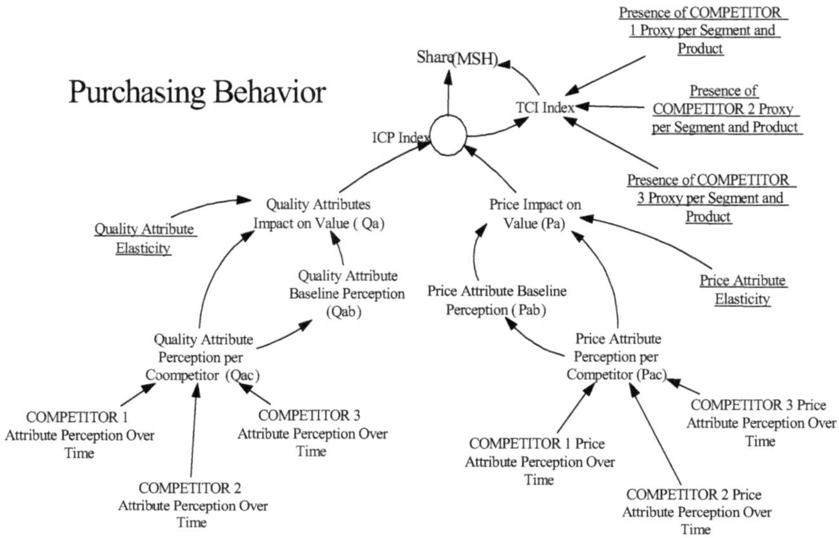

Figure 7.2. Original team design of the purchasing behaviour model

The leverage over time from successful product improvements is shown by the increasing slope of growth curves over time, typically in the shape of an “S” curve, ramping from accumulating assets and then tapering off from the effects of diminishing returns.

7.5 Modelling Marketshare, Revenue, Gross and Net Operating Profit

How does a product and market strategy impact business revenue? How is revenue over time linked to the product’s price attributes and profit? To answer these questions, we will set out the variable equations formalisation process, after first describing the notations and definition of the main financial model variables:

Subscripts:

$j= 1,...,N$ Competitors, including this business
$s= 1,...,S$ Market segments by shared customer purchase priorities (as available)

Input: Segmented market data

Tc_t	Total potential unit sales in t
Ss^s_t	Size (% of the Tc_t) of the segment s in t (note that this is not a model of building and creating a market or individual segments, but of capturing and holding segment share within the strategic market as it grows or shrinks over time, by these exogenous values)
Tcs^s_t	Total potential customers of the segment s in t
$S_j{}^s_t$	Unit sales of competitor j per segment s in t

Input: Business financial targets /history allocated to this solution product

Sd_{jt}	Standard discount (% of list price) of competitor j in t
Md_{jt}	Margin discount (% of list price) of competitor j in t
$Mt_j{}^s_t$	Market share (weighted by segment) trend of competitor j in period t
LP_{jt}	Competitor j list price in t
LPi_{jt}	Competitor j list price increase in period t
LPd_{jt}	Competitor j list price decrease in period t

Calculations: Solution revenue

R^j_t	Revenue of competitor j in period t

Input: Cost ratios

SGA^j_t	Selling, general and administrative expenses of competitor j in period t
C^j_t	Cost of sales of competitor j, in period t
T^j_t	Taxes of competitor j, in period t

Output: Bottom-line for operations and product planners

GP^j_t	Gross profit of competitor j, in period t

Calculation and output: Bottom line for financial planners

NOP^j_t	Net operating profit of competitor j, in period t
COS^j	Cost of sales factor for competitor j as a percent of revenue
TAX	Tax factor as a percent of net operating profit

Nonfinancial measures will now be used as drivers of financial performance indicators, which is an assumption considered in many examples of current research in this area. For instance, Ittner and Larcker [4] have shown how for 2,491 customers of telecommunications firms, customer satisfaction indexes could be correlated to revenue levels, retention and revenue changes of the firms over time. They conclude that their results offer qualified support for recent moves to include customer satisfaction indicators in internal performance measurement systems and compensation plans.

The main equation links market share to revenue and profit by reproducing a pro-forma income statement of the business (in the equations, the competitor index "*j*" is included in order to maintain the ability to track more than one competitor financials according to the model possibilities).

All of the businesses were dynamic models that, when used, used required pro forma statements to show associated market share, with as much back up information about target segments as possible – either as a percentage goal to be achieved over time that has been set by corporate, or as the assumed result of the planned operational targets tied to business projections.

In addition, working with business controllers led the modelling team to incorporate sales discounts for channel incentives (Sd_{jt} & Md_{jt}), cost of sales (COS^j) and tax (TAX) factors, extending operations targets for individual programs to show front end investments and contribution to shareholder metrics.

To meet corporate planning guidelines, the business case usually had to project market share, revenue, and profit metrics, with details for the next four quarters and summary data over three years.

Once the unit sales per segment are calculated in Equations 7.8 and 7.9, Equations 7.10 to 7.14 formalise the income statement of the business as follows:

$$Tcs^{s}_{t} = Tc_t \times Ss^{s}_{t} \tag{7.8}$$

$$S_j{}^{s}_{t} = Tcs_t \times MSH_j{}^{s}_{t} \tag{7.9}$$

$$R^j_t = \sum_{s=1}^{S} S_j{}^{s}_{t} \times LP_{jt} \times (1-(Sd_{jt}+Md_{jt})) \tag{7.10}$$

$$C^j_t = R^j_t \times COS^j \tag{7.11}$$

$$GP^j_t = R^j_t - (COS^j_t + SGA^j_t) \tag{7.12}$$

$$T^j_t = GP^j_t \times TAX \tag{7.13}$$

$$NOP^j_t = GP^j_t - T^j_t \tag{7.14}$$

Experience in different case studies showed that business planners were conceiving the financial model as shown in Figure 7.3 (where the business net operating profit is derived from the market share). Notice how in this Figure the list price strategy is influenced by the market share trend of the business. For example, as a matter of pricing policy, a constraint was inserted in one scenario that raised or lowered the list price if market share projections fit defined gain or loss criteria.

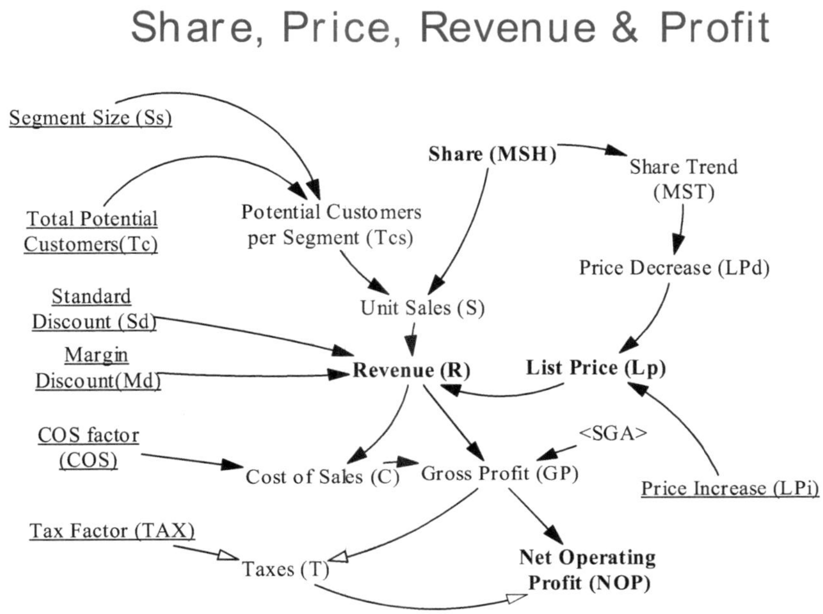

Figure 7.3. Original team design of the financial model

7.6 Modelling Profit Contribution Growth

Once the income statement of the business was modelled and formalised, the modelling team tried to solve the following questions:

- How do we set up a policy to determine the rate of spending we can accomplish?
- What variables should drive decisions about continuing or changing program investments?

The modelling team learned that planning and tracking targets throughout the fiscal year could mean calculating the rate of investment that the business should direct toward a given market opportunity in order to reach its profit goals. This process was however not obvious for many of the members of the model development team, and required detailed conceptualisation and formalisation.

Revising the process followed by the team in order to answer these questions, the notations and definition of the main financial model variables can be described as follows:

Subscripts:

$j = 1,\ldots,N$ Competitors, including this business

Calculations: Changes in financial variable values

Rg^j_t Revenue growth of competitor *j* in period *t*
$SGAg^j$ Growth of selling and general administrative expenses of competitor *j* in period *t*
Cg^j_t Cost of good sold growth of competitor *j* in period *t*
Tg^j_t Taxes growth of competitor *j* in period *t*
PC^j_t Profit contribution of competitor *j* in period *t*
ICF^j Investments constraint factor in competitor *j*

An important aspect to take into account here is that when channel strategy requires incentives in the form of discounts and payments, these costs are added to the computation of net sales as a deduction to compute revenue.

In the example scenario that follows, an existing product is considered, and therefore it is assumed that conditions to increase investment map closely to changes in the financial variable values.

The following Equations (7.15 to 7.18) represent value changes in growth in revenue, cost of sales, SGA expenses, and taxes.

$$Rg^j_t = R^j_t - R^j_{t-1} \tag{7.15}$$

$$Cg^j_t = C^j_t - C^j_{t-1} \tag{7.16}$$

$$SGAg^j_t = SGA^j_t - SGA^j_{t-1} \tag{7.17}$$

$$Tg_t = T^j_t - T^j_{t-1} \tag{7.18}$$

Profit contribution growth is defined as the difference between projected revenue growth, and the sum of the accumulated growth in the other three variables (see Equation 7.19).

$$PCg^j_t = Rg^j_t - (Cg^j_t + SGAg^j_t + Tg_t) \tag{7.19}$$

Finally, SGA expenses for the following year are calculated by considering the profit contribution, revenue growth and other factors. To illustrate how this is done, let us look at the example, based on a real case, presented in Table 7.1.

Here we show how the model can be used to set target spending levels by mapping the pro-forma statement ratios, the proposed spending to increase specific attributes, and the expected returns from a strategy specifically engineered to influence a target segment.

In the example in Table 7.1, profit contribution of Company X, in year 2, could be calculated as follows:

Table 7.1. Numerical example for the determination of investments in the model

	Company X, profit and loss statement					
	YR1		YR2			
	K$	% Rev	K$	% Rev	Growth $	Growth%
Sales	2000.00		3000.00		1000.00	
Standard discount	840.00	42%	1,260.00	42%	420.00	
Margin discount	60.00	3%	90.00	3%	30.00	
Revenue	**1100.00**	**100%**	**1650.00**	**100%**	**550.00**	**50%**
Cost of sales	825.00	75%	1,320.00	80%	495.00	60%
Gross profit	275.00	25%	330.00	20%	50.00	18%
SGA	**44.00**	**4%**	**50.00**	**3%**	**6.00**	**14%**
Net operating profit before taxes	231.00	21%	280.00	17%	50.00	22%
Tax factor	69.00	6%	84.00	5%	15.00	22%
Net operating profit after taxes	162.00	15%	196.00	12%	35.00	22%

$$PCg_{YR2} = Rg_{YR2} - (Cg_{YR2} + SGAg_{YR2} + Tg_{YR2}) = 550 - (495+6+15) = 35 \text{ K\$} > 0$$

In Company X, growth in profit contribution is therefore positive, and revenue growth (in %) is more than three times *SGA* growth (in %) during the last year (50% >14%). This seems to be an optimal proportion for Company X to increase its spending. Suppose, for instance, that when the above conditions are fulfilled, the company increases *SGA* expenses by half (*ICF*=1/2) of the revenue growth (in%), then SGA_{YR3} would be calculated as follows:

$$SGA_{YR3} = SGA_{YR2}\,((1+ICF(\,Rg^{j}_{YR2}/R^{j}_{YR1}))$$

$$=50(1+0.5(550/1100))=50(1.25)=75 \text{ K\$}.$$

Then, this example would be formalised as a policy constraint in our model as shown in Equation 7.20:

$$SGA^{j}_{t+1} = \begin{cases} SGA^{j}_{t}(1+ICF^{j}(Rg^{j}_{t}/R^{j}_{t-1})), \\ \qquad if\ PCg^{j}_{t}>0 \text{ and } (Rg^{j}_{t}/R^{j}_{t-1})>3\times(SGAg^{j}_{t}/SGA^{j}_{t-1}) \\ SGA^{j}_{t} \quad , Otherwise \end{cases} \tag{7.20}$$

The investment model in this example, drawn from actual planning scenarios, was represented with the planning teams as shown in Figure 7.4. In this diagram, which is the original team design, the reader can appreciate a balance loop that shows how the rate of growth in profit contribution conditions the growth of the *SGA* expenses, while *ICF*, *Rg* and *SGAg*, limit that growth.

Notice again how policy could depend on other variables according to specific business and market conditions. See, for instance, comments about market share in previous Sections.

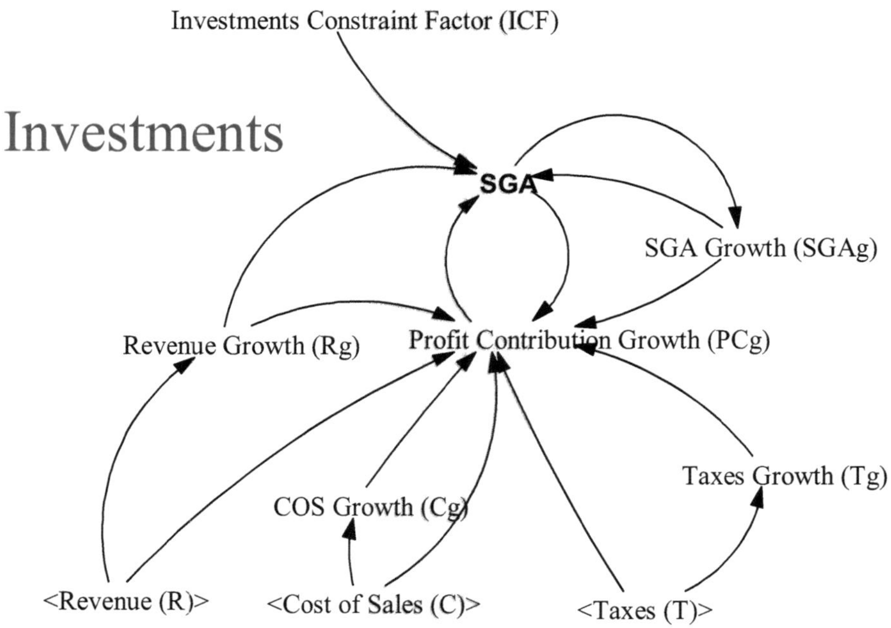

Figure 7.4. Original team design of the investments model

7.7 Transforming a Dynamic Simulation Model into a DSS

Decision Support Systems (DSSs) are tools that an organisation uses to support and enhance decision-making activities [5]. Recently, Decision Support Systems, are becoming a must for any organisation in order to stay competitive and survive in today's dynamic environment.

Early use of decision support analysis were marketing decision support systems (MDSS), defined [6] as a coordinated collection of data, system, tools and technology, with supporting software and hardware by which an organisation gathers and interprets information from business and environment and turns it into a basis for marketing action.

Within the field of marketing, Higby and Farah [7] found that in the US, 32% of the companies have installed some form of marketing DSS (based on a survey among 212 executives); in the Netherlands, Van Campen *et al.* [8] estimated the penetration of decision support systems in marketing at 37% (based on a survey of 525 companies with over 10 employees and marketing manager present in each).

Companies and business planners have recognised the strategic importance of MDSS and are stepping up their investments in information technology for marketing. Adoption of MDSS is higher in companies with consumer products as compared to industrial (business-to-business) products companies, and in companies with more market information available [9].

Their objective is to support a decision making process which is primarily a matter of reasoning (using the mental models of the manager) and analogising (based on stories about similar events retained in mind). For instance, Van Bruggen *et al.* [10] found that managers who use a DSS are less inclined to anchor their decisions on earlier decisions compared with managers who do not use the system. Similarly, these authors found that the incorporation of model-based results into a DSS should be especially beneficial. Prominence effects, overconfidence and other biases are reduced for managers who use model based results DSSs relative to managers who do not. In the literature we find that, although the applicability of some marketing models to real-world problems has been questioned [11], there have been many examples of successful marketing model applications (see for instance [12,13]).

Beyond marketing, another of these models application is within the new products area [14], trying to understand the dynamics between changing demand and the entry and exit behaviours of competitors in the market place. These studies model demand and number of competitors simultaneously, and empirically investigate some hi-tech markets. Further models try to bridge between new product introduction and marketing to understand the relationship between the number of competitors and the rate of technology diffusion [15], or to tie conceptual design in a new product introduction with cost modelling and marketing considerations [16].

In this section we will try to go further to model product design and marketing innovations to anticipate and explain the way collaborative teams, both within firms and between partner businesses, may gain and retain customers in a very competitive hi-tech marketplace. The model also considers the expected response of a changing set of competitors.

The simulation model confirms through team review that we have captured the behaviour that explains their customer segment response to changes in product attributes and price, creating collective understanding of the existing business environment, and also able to be validated by historical data when available. This can be transformed into a DSS model by examining the impact on share, revenue, and profit from engineering and manufacturing changes made to product attributes and prices, as well as changes made to influence the customer's perceptions, given what we believe to be true about the business dynamics.

We will now show an example, case study, of the model used as a framework for a scenario (simulation) where business planners may explore specific product improvement strategies. The simulations calculate expected results in the context of current competitor investment and response, and planners can choose the best strategies to best meet business (financial and operational) targets and forecasts.

Previous Sections showed details within the three sub-models included in Figure 7.1, containing our general model overview. With respect to System Dynamics modelling, we note the importance of causal diagrams. They helped identify feedback mechanisms in the sub-models, to visualise how these could impact the way business grows. The possibility to shift from solely numerical data to a graphical representation provided the opportunity for dialogue and eased mutual understanding, especially for people playing different roles within the business planning process.

The model as shown allows us to study many product and go-to-market scenarios, with enough rigour to focus quickly on key assumptions and to build confidence and consensus when changing business plans.

7.8 Sample DSS and Case Study

7.8.1 Introduction

This is an example of the model's use, based on a real instance where planners considered a possible strategy to improve three product attributes (see Figure 7.5).

Given a stable organisation and product architecture (rare in hi-tech business), the model was first validated with a timespan of 3 years, 1995–1998, for consumer products sold through resellers in a mature market in which the firm was dominant. In this ideal but unusually stable case, with the product attributes that customers hold most important, and the attribute elasticity in each segment, the modelling team was able to replicate the market response to attribute investments during those years with reasonable accuracy.

Using the model for this scenario, planners wanted to know whether or not to continue the same rates of spending increase for the same three attributes (scalable, network friendly and easy to purchase) over the following 3 years, as shown in Figure 7.5, assuming that these would result in the same kind of increase in value perceptions for the said three attributes. They wanted to know what kind of business results they could expect and why, in order to justify the rate of spending they would fund to meet their growth objectives.

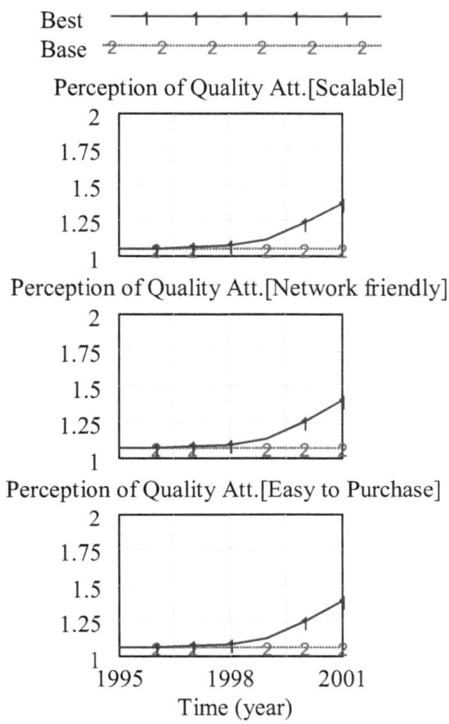

Figure 7.5. Base and best case of attribute improvement

The scenario results shown below in Figure 7.6 illustrate that there is an incremental gain in market share in all three market segments the team was considering (named Soho, Small and Medium), but that the gain is greatest in the segment called "small", where it was possible to reach close to a 5% gain in market share. Segment "Soho" responds very little, and there is only a small gain in the segment "Medium".

In calculating the profit impact, the model uses input from planners about segment growth, share size, competitor strategy, segment elasticity and expected price or cost changes, all occurring at the same time. The planning team, which includes marketing, engineering, finance, supply chain management, and division executives, reviews the scenario output to understand the results, confirm underlying assumptions, and agree where they would redirect spending.

In addition to being able to "drill down" to underlying causes, the scenario shows the net operating profit the business could expect, if those investments are accomplished, given the corresponding yearly increase in *SGA* (see Figure 7.7). The business planners used this outcome analysis to adjust their planned investment program.

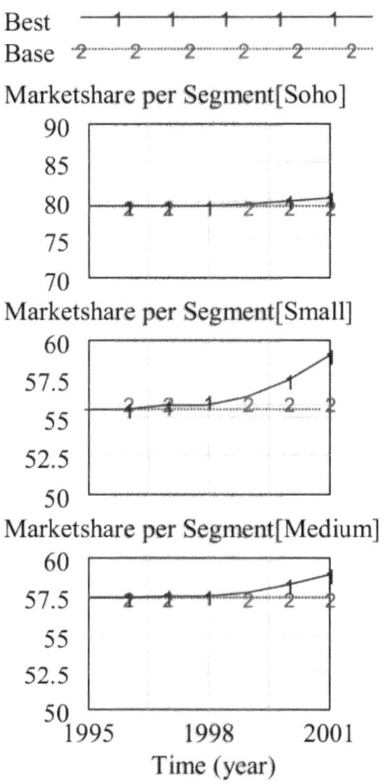

Figure 7.6. Base and best simulations for market share improvement (notice that last two graphs are in a different scale to the first one)

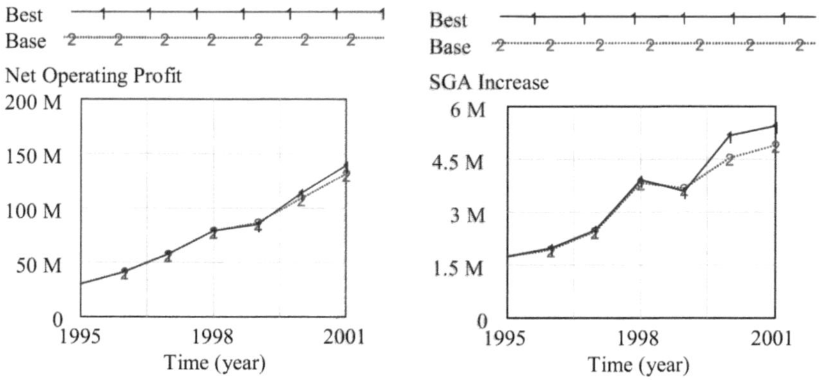

Figure 7.7. Base and best simulations, expected *NOP* (after taxes) and increase in *SGA* per year

7.8.2 From a Simulation Model to a Decision Support System

Now we give an example of how this model can take advantage of optimisation techniques to compare alternative attributes investments, converting the model into a fast Decision Support System.

A modified Powell method is used to carry out these optimisations properly. This is a direct-search numerical optimisation technique which does not need to evaluate the gradient, and which is very suitable for the analysis of dynamics of complex nonlinear control systems. This technique is well known among direct-search methods, being able to derive a very fast convergence.

The basic idea behind Powell's method [17] is to break the N dimensional minimisation down into N separate 1D minimisation problems. Then, for each 1D problem a binary search is implemented to find the local minimum within a given range. Furthermore, on subsequent iterations an estimate is made of the best directions to use for the 1D searches. Some problems, however, are not always assured of optimal solutions because the direction vectors are not always linearly independent. To overcome this, the method was revised [18] by introducing new criteria for formation of linearly independent direction vectors; this revised method is called *The Modified Powell Method.*

In this example, we use the model to select an investment focus for the next 3 years. From the list of attributes that represent the decision factors for solution customers, we compare the impact of improving each attribute by a planned percentage. Each column in Table 7.2 evaluates the choices according to a specific criterion. The table is used one column at a time, with each one representing a different planning scenario.

The table ranks the attributes according to each criterion while in the final rows, the associated percentages tell us how much better the highest level of ranking is than each of the lower level ranks.

In order to optimise their investments, the planners first define the criteria for optimisation to be used in each scenario, the calculation represented here by the column headings. Planners normally consider more than one criterion, and usually include both financial and market-based metrics, representing various business objectives for the planning period. They then define optimisation variables for the ranking calculations (first column). Later, we can ask the model, by using multi-parametric optimisation (considering cumulative evaluation of the payoff), which attributes should be the spending priority to meet best the business criteria shown in the column heading.

The way to read the table is as follows: if we pursue maximising Revenue, first choice for attribute investment should be Reliability and Easy to Purchase; if you increase perception of either of those attributes by a targeted percent (which you assume you will do if you spend according to your plan), your results in terms of revenue will be 7% higher than in Scalability, Network Friendly, or Service Availability, and 14% higher than in any attribute with a 3, and so on.

The model here becomes a powerful and flexible planning tool. You may even explore multiple objectives (see last column that includes both Revenue & Share). The same planning team might go on to explore other investments goals, such as increasing the presence in various segments. This constitutes a Decision Support System that adds clarity and rigor to targets and product/program strategy.

Table 7.2. Example of output from the Decision Support System (not case described above)

	Optimising criteria (equivalent to a different planning scenario)					
	Revenue	Share	Net Profit	Share in Segment x	Share in Product Y	Revenue & Share
Attributes						
Reliability	1	2	6	4	2	1
Easy to purchase	1	2	6	3	2	2
Scalability	2	2	6	3	2	3
Network friendly	2	2	6	3	2	3
Service abailability	2	2	5	3	1	3
Connectivity	3	2	1	2	3	3
Plug and play	4	3	2	4	4	4
...etc.	5	4	4	4	4	4
Relative level of Importance						
(1) more than (2)	7%	14%	20%	13%	12%	5%
(1) more than (3)	14%	20%	60%	15%	20%	16%
(1) more than (4)	78%	+%	260%	+%	+%	+%
(1) more than (5)	545%		340%			
(1) more than (6)			400%			

7.9 Managerial implications

7.9.1 Respond to Market-driven Demand

As we mentioned and characterised in Section 6.1, managers in hi-tech markets face unique challenges. Business planners represent the needs of engineering, marketing, sales, order, delivery, support and service teams. They face changes driven by technological advances, volatile demand, global competition, emerging standards and significant uncertainty about what drives their customer's decisions to buy. This decision support system views the business as a dynamic feedback system to:

1. sense an opportunity matched with an ability to respond to market with value;
2. create value – balancing features and price – and communicate that value to customers in a target segment ;
3. grow with the market, faster than the competition;
4. create early barriers to entry for emerging markets;

5. confidently redirect resources based on changes in customer purchasing behaviour, competitor investment, and the payback that can be expected from the required additional investment.

7.9.2 Segment According to Customer Purchase Priorities

Wherever markets are segmented by customer value and buying behaviour, decision makers may use this model to compare expected financial returns on alternative investments that appeal to some segments more than others. Investments that affect a specific attribute have different implications for each segment, with results for share, revenue and profit that also reflect external changes in size of that targeted segment and of the market demand overall.

Specific investments considered by teams with whom these analyses were done in the past include:

- reseller discounts;
- pricing strategy;
- one-to-one relationship marketing programs;
- advertising to raise target customer awareness;
- new channel development;
- new product and technology introductions;
- introduction of non-branded offerings;
- forward contracts to secure critical part supply; and
- collaborative communication backbones for demand and fulfilment chains.

7.9.3 Focus on the Vertical Dimension of Business Planning

There is only one "product" in the model presented as an example, but in hi-tech sectors, such as telecom infrastructure or medium business manufacturing, the end "product" is a solution, i.e. multiple component products with different cost structures bundled for this market to meet this set of attributes.

Financial targets usually represent product businesses selling into numerous markets, where go-to-market, sales, service, and channel investments are treated as programs, charged with achieving specific market objectives. Although current financial data usually come to the team as product business targets, most critical investment decisions must also consider the impact of changes in attributes and customer perception of value for a solution which will determine its success or failure.

7.9.4 Traction from Precise Go-to-market Strategy

Initiatives to improve business performance are directed toward specific solution attributes:

- quality attributes are improved by investments to improve features, performance, power requirements, footprint size, integration, customisation,

delivery, localisation, scalability, interoperability, quality, channels, and alliances;
- price attributes are improved by investments in aggressive sourcing, parts availability, risk management, order and forecast management, channel incentives, discounts, rebates, advertising, web-based collaborative infrastructure, and synchronised product upgrades.

The critical assumption – that your planned spending will indeed increase customer perception and impact sales as you expect – needs to be confirmed as quickly as possible.

In addition to mining existing market research, planners working with the model development team tended to gain confidence through immediate action guided by the decision support system, with a rapid "pilot", limited in scope and carefully observed to measure and confirm perception and response. Thus, for today's hi-tech businesses, strategy and tactics tend to merge, each informing the other in a rapid exchange between precise action and useful learning.

7.10 Conclusions and Further Research

In this chapter, and with the previous case study, there is a detailed description of how simulation models can be used to support product and marketing investment decisions.

How to derive a DSS from a high level model structure was described and was later formalised in previous chapters. It was shown how the model can be used in business planning to explore a very specific problem, and we have given one example of the model's value as an "engine" of a Decision Support System.

The Decision Support System defined takes into account the horizontal and vertical metrics that together define success for current hi-tech businesses, matching each investment strategy to specific attributes of customer value and business results. Concurrently, the model structure defined allows incorporating other important characteristics of hi-tech markets that are just emerging but will soon be relevant factors in business investments decisions.

System Dynamics simulations greatly improve analysis of go-to-market strategies, integrating customer knowledge with simulations to analyse spending trade-offs in features, services, support, integration, channel incentives, pricing, and advertising.

The payback over time is shown in the output from this formal System Dynamics model, a powerful DSS tool offering the opportunity to compare strategies for a segmented market, under different scenarios, with customised metrics.

7.11 References

[1] Cleland AS, Bruno AV, (1996) The Market Value Process. San Francisco: Jossey-Bass Publishers.

[2] Ross J, Georgoff D, (1991) A survey of productive and quality issues in manufacturing. The state of the industry, Industrial Management, 3(5): 22–25.
[3] Carpenter GS, Cooper LG, Hanssens DM, Midgley DF, (1998) Modelling asymmetric competition. Marketing Science, 7(4): 393–412.
[4] Ittner CD, Larcker DF, (1998) Are nonfinancial measures leading indicators of financial performance? An analysis of customer satisfaction, Journal of Accounting Research, 36 (supplement): 1–31.
[5] Bhatt GD, Zaveri J, (2002) The enabling role of decision support systems in organisational learning. Decision Support Systems, 32(3): 297–309.
[6] Little JDC, (1979) Decision support systems for marketing managers. Journal of Marketing, 43: 9–26.
[7] Higby MA, Farah BN, (1991) The status of marketing information systems, decision support systems and expert systems in the marketing function of U.S. firms. Information and Management, 20: 29–35.
[8] Van Campen PAFM, Huizingh KRE, Oude Ophuis PAM, Wierenga B, (1991) Marketing Decision Support Systemen bij Nedelandse Bedrijven. Eburon. Delft.
[9] Wierenga B, Oude Ophuis PAM, (1997) Marketing decision support systems: Adoption, use and satisfaction. International Journal of Research in Marketing, 14: 275–290.
[10] Van Bruggen GH, Smidts A, Wierenga B, (1998) Improving decision making by means of marketing decision support system. Management Science, 44: 645–658.
[11] Simon H, (1994) Marketing science´s pilgrimage to the ivory tower. In: Laurent, G., Lilien, GL, Preas, B. (Eds.). Research Traditions in Marketing. Boston: Kluwer Academic Publishing.
[12] Little JDC, Lodish LM, Hauser JR, Urban GL, (1994) Commentary, in: G. Laurent, G.L. Lilien, B. Pras (Eds.), Research Traditions in Marketing, pp.44–51. Boston: Kluwer Academic Publishing.
[13] Parsons LJ, Gijsbrechts E, Leeflang PSH, Wittink DR, (1994) Commentary, in: Laurent G, Lilien GL, Pras B (Eds.), Research Traditions in Marketing, pp. 52–78. Boston: Kluwer Academic Publishing.
[14] Kim N, Bridge E, Srivastava RK, (1999) A simultaneous model for innovative product category sales diffusion and competitive dynamics. International Journal of Research in Marketing, 16: 95–112.
[15] Bridges E, Ensor KB, Thomson JR, (1992) Marketplace competition in the personal computer industry. Decision Science, 23(2): 467–477.
[16] Vollerthun A, (2002) Design-to-Market: Integrating conceptual design and marketing. Systems Engineering, 5(4): 315–326.
[17] Powell MJD, (1964) An efficient method for finding the minimum of a function of several variables without calculating derivatives. Computer Journal, 7(2): 155–62.
[18] Powell MJD, (1968) On the calculation of orthogonal vectors. Computer Journal, 11(2): 302–304.

Part III

Modelling Back-end Issues In SCM

8

Back-end Issues Related to Supplier Management

8.1 Contract Structures for Supplier Management

In this part of the book we will define, characterise and simulate different generic types of supplier contracts to accomplish varying degrees of security and flexibility. We will then simulate a portfolio of these contracts applied to secure a single part, with the purpose of extending and refining portfolio valuation. We will focus our attention on business dynamics based on current best practices in portfolio management, as evidenced by leaders in volatile technology businesses.

The strategic part procurement system, as modelled here, includes in-transit and warehoused inventories to accommodate demand variability as well as asynchronous production and shipping lead times. In addition, the portfolio structure itself periodically creates excess inventory as negotiated terms and policies are executed in various conditions. Given these inventories and related costs, we will show how optimisation techniques can be used to measure the tradeoffs between alternative portfolio structures.

Finally, we will describe how this more complete valuation of the portfolio is critical to option pricing models, with the ultimate potential for indexing as seen in other commodity markets.

8.2 Competitive Prourement Strategies: Global and Multiple Sourcing

The use of multiple sourcing is assumed to diminish the risk of delays or failure on the part of just one supplier [1], and may also encourage their performance as regards delivery and quality [2]. Other factors influencing multiple sources are economics, geography, organisational policy and buyer inertia. Multiple sourcing should be adopted as a procurement strategy in those cases where items are critical in the production process and which incur high cost if the production lines are stopped [3]. Where supplier contracts are structured on volume discounts, higher part prices might be charged when demand volume is allocated to multiple sources, but that increase can be seen as an insurance against the higher total cost of stopped production [4].

Establishing relationships with different suppliers is not an easy job though; Ellram [5] offered a normative model to guide the process of developing and implementing partnerships, consisting of four phases:

- Phase 1 is defined as the preliminary phase of establishing the strategic need for a partnership, forming an internal team, and ensuring the top management support for developing a partnership. During this preliminary phase, procurement internal teams have to develop tools to demonstrate the need of partnerships, to value them, to obtain the maximum support from the top management for their development.
- Phase 2 is to determine the selection criteria for potential partners.
- Phase 3 consists of the screening and assessment of candidates (considering aspects such as cultural compatibility, long-term strategies, financial stability, management compatibility, and location of facilities).
- In Phase 4, expectations of the relationship are established, and suitable monitoring and feedback mechanism are put in place. Long-term partnership with the suppliers means an ongoing relationship between firms involving commitment over an extended time period, and a mutual sharing of information and the risk and reward of the relationship [6]. The buyer may expect to reduce the cost of the purchase, which could be critical to ensure product margin, and secure a reliable source, which is very important in times of scarcity [7]. Long term partnerships with suppliers may also produce clear benefits when the buyer is delivering solutions for a lasting hi-tech infrastructure component, with lengthy and expensive certification procedures (e.g. the telecom industry, where changes to the components or design of these products is problematic).
- Finally, in Phase 5, the relationship is continually evaluated and adjusted.

8.3 Types of Contractual Relationships with Suppliers

The terms and conditions of the individual contracts that the buyer will maintain with suppliers act together to cover forecasted part demand. Setting up a set of contracts with suppliers for a shared commodity part configures what we call a procurement contract portfolio. The different portfolio configurations must be considered in the context of the volatility in world commodity prices and combined product demand forecasts.

Partnership with suppliers should mitigate the different dimensions of risk: demand, price and source concentration (geographical) risk.

Different strategy options for supplier contracts portfolio will face this problem, allowing the company to gain market and price positions, to ensure required customer service levels, and to diversify and secure source in times of scarcity.

After the involvement in several procurement risk management projects, Table 8.1 presents a vision of the most popular contract types found within the hi-tech industry, their purpose, and the way in which the relationships are established between the parties.

Table 8.1. Contract types, demand forecast certainty, their nature and benefits

Contract type	Demand Tier	Nature of the relationship	Purchasing objectives
Structured contracts including firm commitments (S Contracts)	Tier 1 Certain demand forecast	– Cooperative and collaborative buyer seller relationship. Expectations of continued duration – Shared responsibilities for task to reach quantity, price and delivery terms	– Improved quality, reliability and total system operational cost – Lower investment risk – Increased demand and price certainty
Flexible quantity and price contracts (F Contracts)	Tier 2 Upside demand forecast	– Firms commitments to supply and purchase within a specified volume and pricing range – Normally includes a prepayment or monetary agreement in return for committed volume and pricing terms	– Sharing price risk and opportunities – More predictable earning streams – Buyer is better able manage demand risk – Supplier obtains better pricing and opportunities to grow business – Better production and investment planning
Short or Specific term / traditional contract (ST Contracts)	Tier 3 Unlikely but possible demand forecast	– Transactional, discrete or short term events – Based on alternative sourcing negotiated periodically	– Right volume at the right time with the right price

8.4 Procurement Risk Management at HP. A Case Study

This case study is adapted from a presentation of the Hewlett Packard Procurement Risk Management Group (PRMG) to the Council of Supply Chain Management Professionals and describes HP's procurement risk management approach and initiatives during the years 2000–2005 [8,9]. The idea of this case study is to contextualise somehow the dynamic modelling work that will be presented in the following chapter, as a tool to deal with these types of procurement issues and problems.

"...In 1999–2000, HP faced significant price increases and an availability shortfall for flash memory used in highly profitable printer lines. Demand for flash memory grew exponentially due to increasing demand from cell phone manufacturers, and an expected shortfall in flash memory threatened printer shipments. To assure future availability of flash memory, and protect HP's printer profits, HP decided to enter into a binding long-term contract with a major flash memory supplier. The uncertainty in the future price and availability of flash memory, and HP's own demand uncertainty for this type of memory, made specifying the terms and conditions of the contract very difficult.

More precisely, HP had to evaluate the following aspects in order to avoid a risky and imprudent long-term commitment:

1. what to pay for flash memory over the next few years and how to structure payments;
2. how much to buy and how to structure delivery terms;
3. how long a horizon should the contract cover and when is the best time to sign the agreement;
4. provisions to include to secure compliance..."

Nagali et al. [8] describe how the Procurement Risk Management (PRM) program was launched at HP in August of 2000. The initial idea was to develop and standardise methods to improve decisions when managing suppliers' risk. Risk management has later become a critical strategy for procurement at HP.

8.4.1 Procurement Uncertainties

Procurement risks are a consequence of uncertainty in [9]

- product demand;
- component price; and
- component availability.

Many manufacturers know these risks very well. For instance Ford posted a $1 billion loss on precious metals inventory and forward contract agreements in December 2001. Demand uncertainty has also caused cycles of product and component shortages followed by inventory build-ups and write-downs. For instance Cisco took a $2.5 billion inventory write-down in April of 2001 due to weakening demand for networking products.

The hi-tech components that HP purchases can exhibit even more volatility. In the case of Ford, palladium prices doubled over the year 2000 and decreased by over 50% in 2001. By comparison, the price of DRAM memory used by HP dropped by over 90% in 2001 and more than tripled in 2002 [9].

Availability of hi-tech components can be uncertain in periods of high demand as a consequence of suppliers' allocation policies. This may lead to supply and delivery disruptions (see Figure 8.1), such as those created by environmental factors (e.g. the earthquake in Taiwan in late 1999). See other dated journal news examples in Figure 8.2.

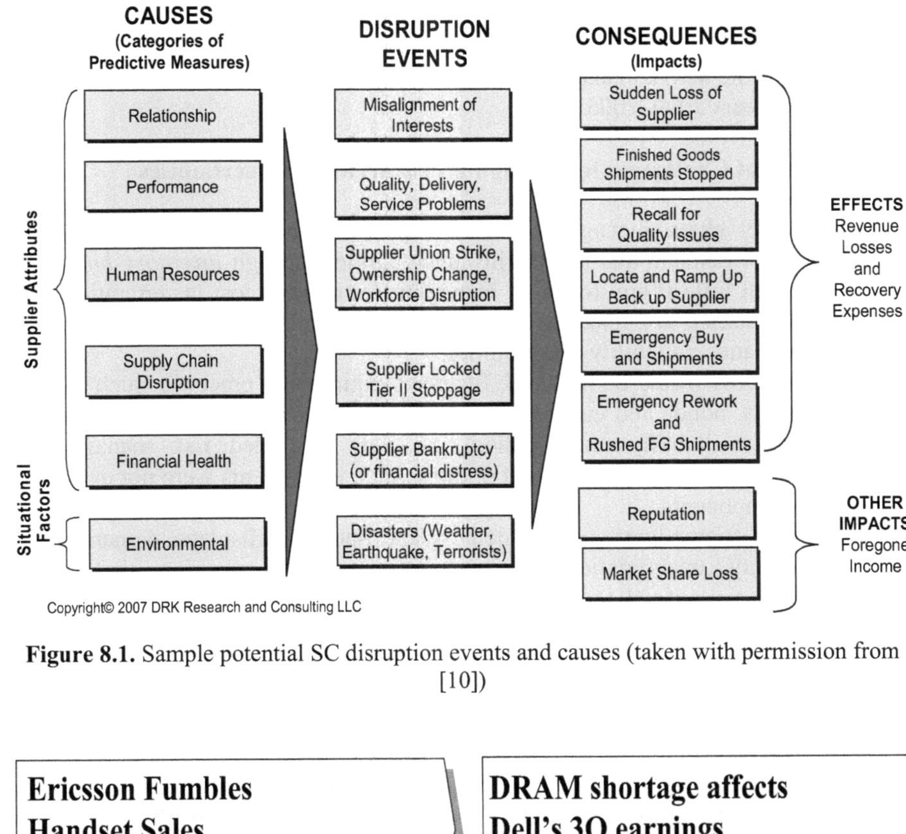

Figure 8.1. Sample potential SC disruption events and causes (taken with permission from [10])

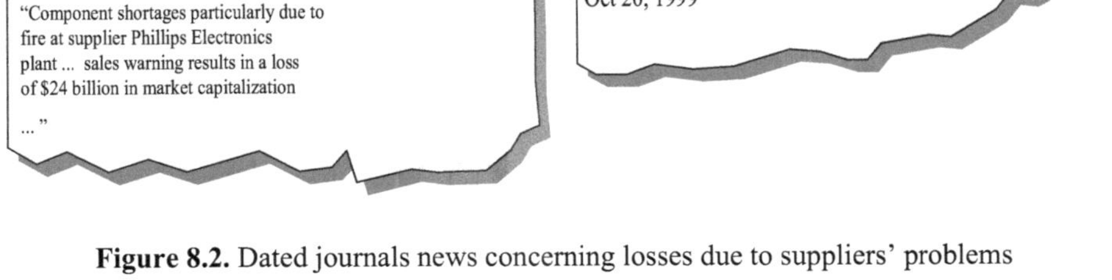

Figure 8.2. Dated journals news concerning losses due to suppliers' problems

In mid-2000, HP signed the long-term binding contract with a major supplier to manage actively the substantial future price and availability uncertainty of flash memory. There were significant 'incremental' risks to HP in entering into the forward contract. For example, if HP demand weakened, then committing to buy a fixed quantity would result in significant inventory build-ups and write-offs. If flash memory prices dropped, HP would pay more through the fixed-price commitment than its competitors.

To ensure minimal risks due to the forward contract a quantitative framework was develop to compare in detail the long-term demand, price and availability uncertainty scenarios for flash memory, and compare those to the quantity and price HP committed to in the contract.

The long-term binding contract for flash memory signed in mid-2000 thus set the course for the active management of procurement uncertainties and risks at HP and over the years 2000–2005.

8.4.2 Technical Challenges in Managing Procurement Uncertainties

Current supply chain management (SCM) practices by 2000 emphasised the management of demand and availability uncertainties through *inventory buffering strategies*, with little if any focus on managing component cost uncertainties. At that time, HP started to pursue the simultaneous measurement and management of demand, cost, and availability uncertainties.

Different tools would be required: financial engineering practices, such as that used for stocks, bonds and currency (management of cost uncertainty though not demand and availability uncertainties), as well as traded risk management instruments, for example, call and put options; such instruments were not available for hi-tech components.

For hi-tech components such as memory and flat-panel displays, demand, cost and availability uncertainties are equally important, requiring that these uncertainties be managed together [9]. Existing supply chain management and financial engineering practices could not, by year 2000, be directly applied to the management of procurement risks of hi-tech components.

As a result HP invented the PRM framework that enables the simultaneous measurement and management of demand, cost and availability uncertainties. HP also developed proprietary software tools to support PRM, since existing SCM and ERP software based on the current SCM theory did not support risk management.

HP's procurement risk management approach involves two steps [8]:

1) measuring uncertainties associated with buying commodities; and
2) managing these risks using structured contracts.

8.4.3 Measuring Uncertainty. The Scenario Approach

The first step in this process is always to estimate uncertainties. Component demand, price and availability uncertainty over time are captured using forecast scenarios. Typically, each of these uncertainties is represented by high, base and low scenarios over time, along with a likelihood estimate for each scenario, as shown in Figure 8.3. Current forecasting approach at most companies emphasise "point" forecast, which in the PRM approach is represented by the base scenario.

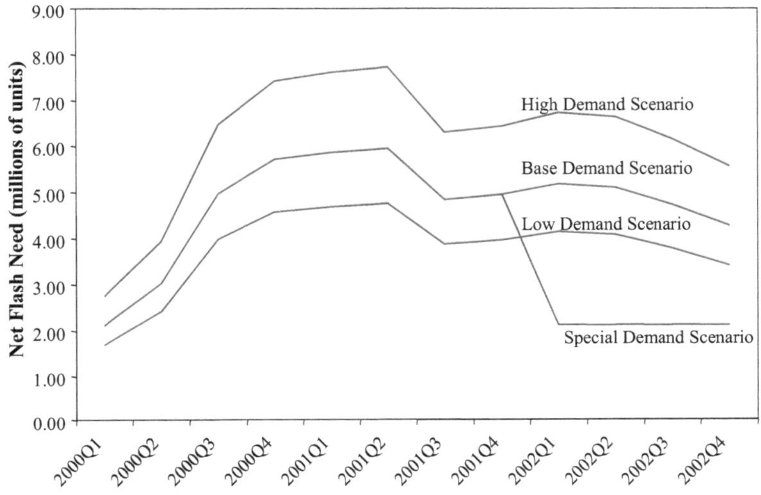

Figure 8.3. Sample scenarios used to quantify uncertainty in component demand

As illustrated in Figure 8.3 for flash demand forecasts high, base and low scenarios are typically defined as the 90th, 50th and 10th percentiles, respectively, of the uncertainty distribution for demand, cost and availability.

Uncertainty around the base scenario is captured by the low and high scenarios. An estimate of the correlation between demand, price and availability uncertainties is also captured. Once the procurement uncertainties are modelled, the current procurement strategy can be analysed to measure the procurements risks involved.

8.4.4 Managing Risks. Structuring Contracts with Suppliers

Once the demand, cost and availability uncertainties are quantified using scenarios, the risks associated with these uncertainties are managed using structured contracts with suppliers. Structured contracts are binding commitments between HP and the supplier, with complex combination of quantity and pricing terms. Quantity terms include fixed and flexible quantity contracts, and percent of total-available-market (TAM); pricing terms include discount-off of market price, fixed price, price-caps and price-floors.

Estimating demand uncertainty illustrates the HP PRM approach and has significant benefits of its own. By quantifying the uncertainty, demand can be segmented according to its risk, as shown in Figure 8.4. Once completed, management teams look for low cost, efficient means to satisfy "certain" demand, and more flexible methods to satisfy likely but "less certain" demand. Approached in this way, the risk and cost to supply products and services is reduced. This means less inventory, less inventory risk, better service levels to customers, lower costs and more competitive product pricing. HP supply chain teams estimate that total supply chain costs could be lowered by as much as 20% by this approach [8].

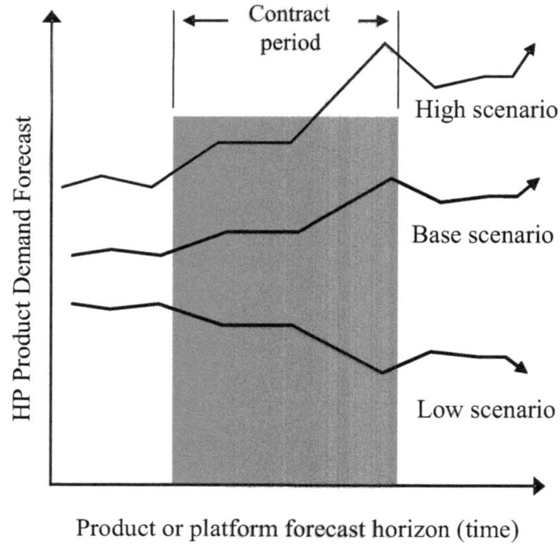

Figure 8.4. Contracts to manage demand uncertainty

For buyers and commodity managers, segmenting demand allows new opportunities for creative contracting with suppliers, in three ways:

- For certain demand, minimum quantity forward contracts assure supply at very low prices since suppliers are often willing to discount for firm quantity commitments because it allows them to manage capacity more efficiently. Committed volumes can be scheduled during non-peak times, and inventory carries no risk. On high volume deals, supplies can modify fabrication lines to significantly reduce costs. According to Nagali et al. [8], one HP printer product team could secure more than 15% discount on committed forward contract in addition to volume discounts! The supplier could modify a conventional process based on HP's binding, forward commitment.
- Less certain demand is satisfied through flexible quantity agreements. Flexible agreements are the most common supplier arrangement in hi-tech industry, so creative modifications of these agreements are usually easy to pull together with suppliers. Fabrication suppliers provide pricing discounts for committed "upside volumes", especially when the volumes are high with potential to grow. Discounts often increase as more volume is purchased. Making these commitments binding eliminates supply risk and provides further cost savings. A significant percent of HP's memory requirements are met through these binding but flexible agreements. Contract horizons generally match HP's product lifecycles time and/or

supplier capacity lead-times. The longer the horizon, the deeper the price discounts and the more binding the supply commitments [8].

- Demand that is least likely to happen can often be satisfied through the open or spot market. As these sources dry up, secondary-sourcing options can be used such as brokers, auctions and product recycling programs. As Nagali et al. mentioned [8]: "These approaches mean higher prices, but are often a better solution than carrying inventory. Furthermore, the supply risk associated with these approaches is often less than expected. HP's customer support teams realised significant inventory savings by recovering critical parts from unsold products. They also found consistent supply for low volume microprocessor demand through auctions, saving on inventories subject to severe price erosion".

8.4.5 The PRM Business Process

HP has developed a PRM business process which defines and coordinates the roles and responsibilities of procurement, planning, supply chain operations, finance and marketing [9]. HP's risk management process is shown in Figure 8.5. Strategy and governance for a particular commodity typically includes approving procurement objectives, establishing metrics and reviewing performance of any existing portfolio of deals.

The deal origination process guides the design of structured contracts to meet procurement objectives and to manage current product and component market uncertainties. For company-wide contracts, commodity managers specify contract terms (originate) that satisfy specific product or divisional objectives, and then a team of commodity managers integrate these specifications into a single contract to leverage purchasing power. These contracts are evaluated during the Contract Evaluation phase to determine their future performance against objectives under various conditions of demand, price and supply uncertainty. Specific attention is paid to situations under which a contract would perform worse than buying without binding terms and conditions. Different software tools are used to evaluate the performance of on going negotiations of structured contracts. These tools are also used for management, review and approval of the contracts. Once approved, these contracts are negotiated and executed using traditional or eSourcing methodologies.

According to Nagali et al. [9], the contract monitoring process guides:

1. the backward-looking measurement of HP's and supplier's performance against commitments made in the structured contracts;
2. the determination of the past performance of a structure contract (or combination of contracts) when compared to previously established metrics;
3. future performance of an existing portfolio of contracts under changed forecast scenarios for demand, price and availability.

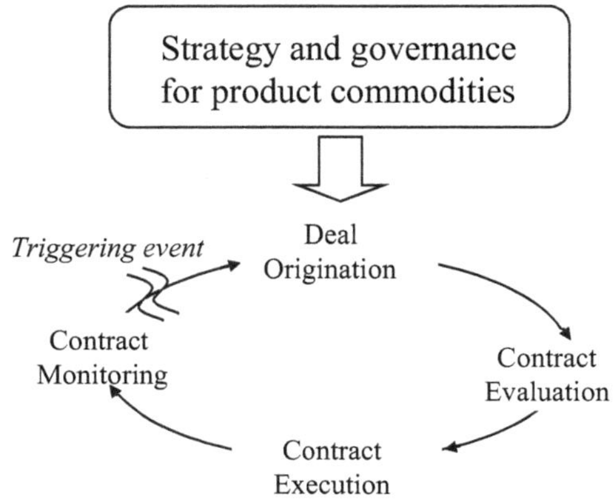

Figure 8.5. HP's procurement risk management process

8.4.6 Benefits from Implementing PRM at HP

Over the years 2000–2005, HP's declared benefits of implementing PRM include [9]:

1. Material costs savings: PRM deals with quantity commitments lower the supplier demand risks, while also enabling the supplier to cut costs through more efficient planning and production processes. The suppliers share some of this value with HP by the way of discounts on material costs. Through such PRM quantity commitments, HP has obtained incremental material cost discounts up to 5% for standard components, and an even higher discount for custom components, indirect and services procurement.
2. Cost predictability: PRM deals with specific pricing terms enabling HP to manage proactively cost uncertainty. A significant portion of memory is procured using PRM deals thus enabling HP to obtain cost predictability required to protect margin on large customer deals.
3. Assurance of supply (AoS): managing component demand and availability uncertainties is a key objective for PRM at HP. PRM deals have improved AoS for several commodities even under conditions of an industry-wide shortage. For example, nearly a year ago there was an industry-wide shortage for memory, but the PRM deals executed by a particular HP business unit ensured that they obtained 100% of their demand from the suppliers.
4. Inventory cost reductions: the precise measurement of demand uncertainty using PRM software enables HP to optimise inventory levels internally and externally at supplier sites. Such optimisation has cut inventory driven costs by several percentage points for commodities implementing the PRM framework.

8.5 References

[1] Min H, Galle WP, (1991) International purchasing strategies of multinational US firms. International Journal of Purchasing and Materials Management, Summer: 9–18.
[2] Carr CH, Truesdale TA, (1992). Lessons from Nissan's British suppliers. International Journal of Operation and Production Management, 12(2): 55.
[3] Quayle M, (1998) Industrial procurement: Factors affecting sourcing decisions. European Journal of Purchasing and Supply Management, 4: 199–205.
[4] Asanuma B, (1985) The organisation of parts supply in the Japanese automotive industry. Japanese Economic Studies, 15: 32–53.
[5] Ellram LM, (1991) A managerial guideline for the development and implementation of purchasing partnerships. International Journal of Purchasing and Materials Management, 27(3): 2–8.
[6] Ellram LM, Krause DR, (1994) Supplier partnership in manufacturing versus non-manufacturing firms. The International Journal of Logistics Management, 5(1): 45–53.
[7] Virolainen VM, (1998) A survey of procurement strategy development in industrial companies. International Journal of Production Economics 56–57: 677–688.
[8] Nagali V, Hwang J, Sanghera D, Gaskins M, Pridgen M, Thurston T, Mackenroth P, Branvold D, Scholler P, Shoemaker G, (2008) Procurement Risk Management (PRM) at Hewlett-Packard Company. Interfaces, 38(1): 51–60.
[9] Nagali V, Sanghera D, Hwang J, Gaskins M, Baez C, Pridgen M, Mackenroth P, Branvold D, Kuper A, Scholler P, (2005) Procurement Risk Management (PRM) at Hewlett-Packard Company. Presentation at the Council of Supply Chain Management Professionals. (http://www.drkresearch.org/Contact_Us/Risk_Roundtable/HPProcurement.pdf).
[10] http://www.drkresearch.org

9

Modelling a Portfolio of Contracts with Suppliers

9.1 Overview

In this chapter dynamic modelling is used to understand better different issues related to the contractual relationship with suppliers of business strategic parts. As mentioned in Chapter 2, a strategic part is considered as a part that is critical to product success, with global price and availability driven by external market forces that could be sometimes beyond the buyer's control.

Chapter 8 perfectly illustrates this topic; it shows the reader that there is a clear need for action in this hi-tech business area. Organisations within this industrial sector need to develop flexible procurement strategies to deal with this uncertainty.

Dynamic modelling can help to evaluate different strategies and possibilities. In this chapter a variety of tiered contract structures will be defined, characterised and simulated. Attention will be focussed on business dynamics based on current best practices in portfolio management, as evidenced by leaders in volatile technology businesses.

The strategic part procurement system, as modelled here, includes material and information flows to accommodate demand variability as well as asynchronous production and shipping lead times. The portfolio structure may also create excess inventory according to terms and conditions of the contracts structure. Optimisation techniques will then be used to measure the tradeoffs between alternative portfolio structures.

The rest of the chapter is organized as follows. The next Section, 9.2, is a section characterizing the relational and contractual opportunities with suppliers. After that, Section 9.3 is dedicated to the procurement system modelling, with special attention to the modelling of forward and commodity options contracts with suppliers. The selection of a suitable contracts portfolio with suppliers using the model, the managerial implications of the study, and the final conclusions, form the last chapter sections.

9.2 Formal Characterisation of the Contracts with Suppliers in a Dynamic Volatile Business Environment

9.2.1 Notation of the Model Material and Information Flow Variables and Parameters

In order to characterise the relational and contractual opportunities with suppliers, we first present the notation and definition for the main information and material flow variables and parameters of the model:

Information related variables:

Bcp_t Existing backlog of critical parts needs in t
$Busc_t$ Existing backlog of orders in the *nsc* supplier in t
Bfc_t Existing backlog of orders in the *fc* supplier in t
Bsc_t Existing backlog of orders in the *sc* supplier in t
D_t Production line critical parts consumption in period t
$IOusc_t$ *usc* supplier incoming orders in period t
$IOfc_t$ *fc* supplier incoming orders in period t
$IOsc_t$ *sc* supplier incoming orders in period t
Fcp_t Short term critical parts needs forecast in period t
$Fusc_t$ *usc* supplier forecast in period t
Ffc_t *fc* supplier forecast in period t
Fsc_t *sc* supplier forecast in period t
sscp Desired safety period of critical parts in front of the production line
ssusc *usc* supplier safety period,
ssfc *fc* supplier safety period ,
sssc *sc* supplier safety period,

Material flow related variables:

$Iusc_t$ *usc* supplier production rate in period t
Ifc_t *fc* supplier production rate in period t
Isc_t *sc* supplier production rate in period t
$INVhit_t$ Inventory of critical parts in front of the production line in t
$INVusc_t$ Inventory of critical parts on-hand in the *usc* supplier in t
$INVfc_t$ Inventory of critical parts on-hand in the *fc* supplier in t
$INVsc_t$ Inventory of critical parts on-hand in the *sc* supplier in t
$SDhit_t$ Critical parts in hubs or in transit after the suppliers in t
$WIPusc_t$ *usc* supplier work in process in t
$WIPfc_t$ *fc* supplier work in process in t
$WIPsc_t$ *sc* supplier work in process in t
$OSDhit_t$ Units arriving to the inventory in front of the PL in period t
$Ousc_t$ *usc* supplier output from WIP in period t
Ofc_t *fc* supplier output from WIP in period t
Osc_t *sc* supplier output from WIP in period t

Scp_t	Parts that the production line is pulling from the buffer in period *t*
$Susc_t$	*usc* supplier parts shipments in period *t*
Sfc_t	*fc* supplier parts shipments in period *t*
Ssc_t	*sc* supplier parts shipments in period *t*

Material and information flow related parameters:

Lhit	Procurement hubs and in transit lead times
Lusc	Lead time of a supplier with an unstructured contract
Lfc	Lead time of a supplier with a flexible contract
Lsc	Lead time of a supplier with an structured contract
α	Orders forecast smoothing factor
β_S	Fractional adjustment coefficient for the on-hand inventory
β_{SL}	Fractional adjustment coefficient for the work in process inventory

9.2.2 Characterisation of Supplier Contracts in a Volatile Business Environment

In Figure 9.1 we represent a generic demand scenario for parts procurement. Contracts with suppliers are periodically reviewed (*RP*), and each one of them meets a certain level of part supply availability for the buyer. In the same figure, *Dsc* denotes a negotiated constant order and supply flow for S (structured) type of supplier contracts. $Dfc=Dfc_t$ represents an expected, but not certain, upside demand, a variable supply flow arriving through flexible contracts with suppliers.

The reason for *Dsc* to be considered constant is the fact that this would provide the best possible price from the suppliers once they would be able to deduct from price the corresponding inventory holding cost. Moreover, the suppliers would need fewer funds to run the contract with the buyer, and therefore the relationship would be less risky for them.

Finally $Dusc=Dusc_t$ represents the variable supply flow from supplier with demand in excess of levels covered under previously negotiated price and capacity commitments. In cases of high volatility, it may happen that fixed supply flows ($Dsc+minumum\ Dfc_t$) could exceed current part forecast and therefore parts would be accumulated in transit, in hubs or in front of the buyer manufacturing/assembly line. The value of these inventory buffers will be taken into account to determine future orders to suppliers. Notice that in the example of Figure 9.1, minimum Dfc_t, is assumed to be zero.

In Figure 9.2 we represent a generic price scenario for parts procurement presented in Figure 9.1. *Psc* denotes the constant average price for parts supplied within structured flow (S) type of contracts. $Pfc=Pfc_t$ represents the price negotiated for parts bought through flexible contracts with suppliers (F contracts). Finally $Pusc=Pusc_t$ represents the price from supplier unstructured contracts, negotiated only when the need arises.

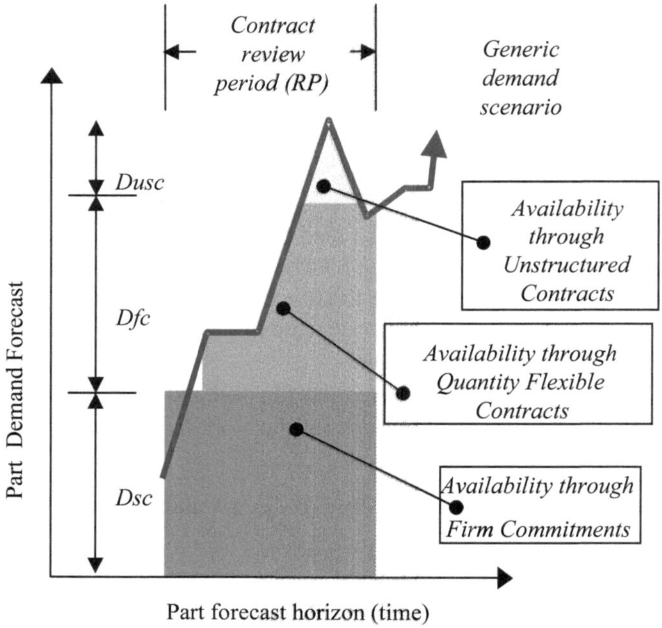

Figure 9.1. Contracts to manage demand uncertainty [1]

For a given time *t,* the fixed price for the structured contracts *Psc* will be known, and we will assume that the price for the flexible quantity contracts Pfc_t will probably depend on the volume discounts, available capacity for the target period, and trends in spot prices (Equation 9.1):

$$Pfc_t = f(p(Dfc_t), Dfc_t) \tag{9.1}$$

Once the contract is arranged, price will be defined (Figure 9.3) and will, in the case of distinct purchase commitments, probably be different. Also, the supplier defines this function *p(t)* in terms of caps and floors of the price increase over the contract horizon (e.g. for purchases commitments over x amount, the maximum expected price increase per quarter would be y% during the contract time horizon, etc.). After this consideration, if the demand cannot be met through the contract portfolio agreements, ($Dt > Dsc + Dfc_t$ for any t), we will have to try to buy parts at the spot price (or very close to that) through unstructured contracts, and the price will be according to market expectations as a function of time $Pusc_t$.

For purposes of risk assessment of operational exposure, it is important to note that for trailing edge technologies, spot prices, although several times higher than negotiated contracts, are often accompanied by severe scarcity – parts are often simply not available in the open market, and temporary line-downs and lost sales must be considered as a possible outcome.

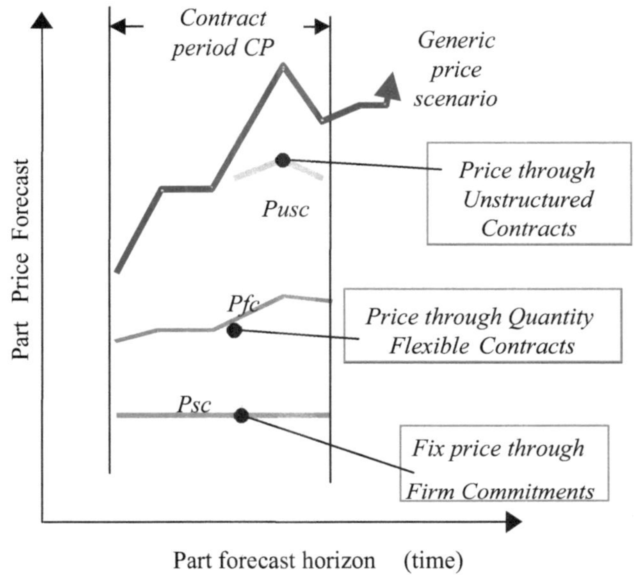

Figure 9.2. Contracts to manage price uncertainty [1]

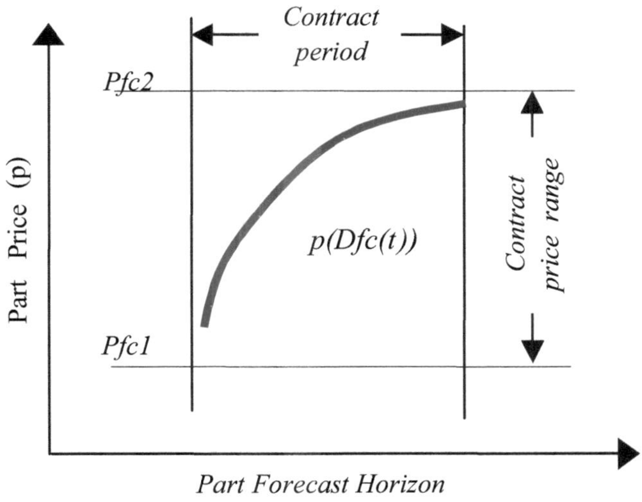

Figure 9.3. The flexible contract formalisation [1]

9.2.3 Modelling the Procurement System. Material and Information Flows

The system for critical parts procurement will now be modelled. It is assumed that the flow of parts will have three possible sources, classified according to the relationship maintained with the suppliers, and shown in Table 8.1.

Every parts flow will have a different price for parts, and lead times are also negotiated as part of the supplier contract. All the flows will go through hubs and transit processes to a buffer inventory, from where the assembly or production lines will pull parts as required (Figure 9.4). The idea behind this is that when a critical part is needed (D_t) in the line, the part is immediately delivered from the buffer inventory (Scp_t). The final critical part inventory buffer ($INVhit_t$) is then maintained to ensure the procurement system service level, but also to accumulate parts in case of demand falling below the level of continuous parts inflow received from suppliers with long terms agreements.

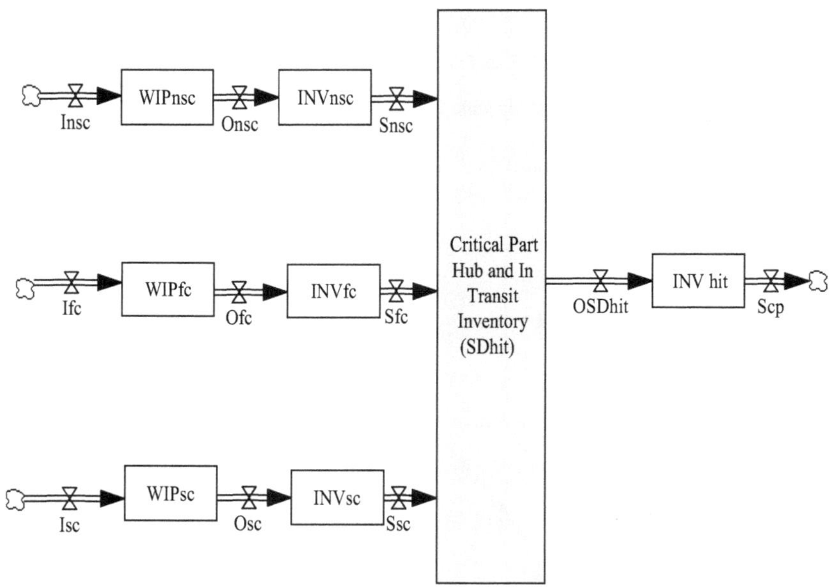

Figure 9.4. The main variables in the entire critical part stock and flow diagram

Equations for the main material and information flow variables will now be presented. These equations are divided in three groups (see Figure 9.4 to locate the variable that is being formalised each time), as follows:

- equations for parts delivered to the production line from the procurement system;
- equations for inventories within the procurement system;
- equations to estimate what is ordered to each supplier.

Equations for parts delivered to the production line from the procurement system:

$$Scp_t = \begin{cases} D_t + Bcp_t, & \text{if} \quad INVhit_t \geq D_t + Bcp_t \\ INVhit_t, & \text{if} \quad INVhit_t < D_t + Bcp_t \end{cases} \tag{9.2}$$

with

$$Bcp_t = Bcp_{t-1} + D_t - Scp_t \tag{9.3}$$

Note that Equation 9.2 shows how backordering is allowed in the procurement system.

Equations for inventories within the procurement system:

Inventories on hand would be:

$$INVhit_t = INVhit_{t-1} + OSDhit_t - Scp_t \tag{9.4}$$

In transit inventory can be determined by:

$$SDhit_t = WIPhit_{t-1} + Susc_t + Sfc_t + Ssc_t - OWIPhit_t \tag{9.5}$$

where Equation 9.6 represents the outputs from the hubs and in transit inventory in time t, equal to the inputs to this inventory at time t-L:

$$OSDhit_t = Susc_{t-L} + Sfc_{t-L} + Ssc_{t-L} \tag{9.6}$$

Equations to estimate what is ordered to each supplier:

In this system it is assumed, for example, that there is a pull system established with flexible suppliers and that the total orders to suppliers (PO_t) are release according to:

1. A forecast such as that in Equation 9.7, with a widely used [2] and popular practice [3] exponential smoothing constant is used. Obviously, this forecast could come directly from the supply chain downstream, assuming reasonable levels of supply chain integration. For the model presented here, we will consider that the procurement teams running the procurement system are elaborating their own forecast though. To choose appropriate values of α, the reader is referred to Makridakis *et al.* [4]:

$$F_t = \alpha D_t + (1-\alpha) F_{t-1} \text{ with } 0 \leq \alpha \leq 1 \tag{9.7}$$

2. An anchoring and adjustment heuristic [5] with fractional coefficients β_S and β_{SL}, once has been shown to apply to this kind of decision-making task [6]:

$$PO_t=Max(F_t+\beta_S(F_t sshit_t - INVhit_t)+\beta_{SL}(F_t Lhit - WIPhit_t),0) \tag{9.8}$$

Then, under these considerations, the final incoming orders received by each supplier can be formalised as follows:

$$IOsc_t= Dsc \tag{9.9}$$

$$IOfc_t= Min(Max(PO_t-IOsc_t,Dfc1_t),Dfc2_t) \tag{9.10}$$

$$IOusc_t= Max(PO_t-IOsc_t-IOfc_t,0) \tag{9.11}$$

For the supplier, under structured contract, we will always order the same *Dsc* units per period.

For the supplier, flexible and unstructured, we will also assume the same kind of decision making process to determine their input production rates and shipments. For instance, the supplier under flexible contract equations could be formalised as follows

Equations for parts delivered to the hub and in transit inventory:

$$Sfc_t=\begin{cases} IOfc_t+Bfc_t, & \text{if} \quad INVfc_t \geq IOfc_t+Bfc_t \\ INVfc_t, & \text{if} \quad INVfc_t < IOfc_t+Bfc_t \end{cases} \tag{9.12}$$

$$Bfc_t = Bfc_{t-1} + IOfc_t - Sfc_t \tag{9.13}$$

Inventories on hand would be

$$INVfc_t= INVfc_{t-1} + OWIPfc_t - Sfc_t \tag{9.14}$$

Supplier output or production completion rate would be

$$Ofc_t= Ifc_{t-L} \tag{9.15}$$

In transit inventory can be determined by

$$WIPfc_t= WIPfc_{t-1} + Ifc_t - Ofc_t \tag{9.16}$$

Orders forecast can then be determined as follows:

$$Ffc_t = \alpha\, IOfc_{t-1} + (1-\alpha)\, Ffc_{t-1} \text{ with } 0 \leq \alpha \leq 1 \tag{9.17}$$

and the supplier production rate is computed using Equation 9.18:

$$Ifc_t = Max(Ffc_t + \beta_S (Ffc_t ssfc_t - INVfc_t) + \beta_{SL} (Ffc_t Lfc - WIPfc_t) ,0) \tag{9.18}$$

In Equation 9.18 we are also assuming there are no delays for the supplier to process their production orders and that their raw material is always available.

Once the equations for the material and information flows have been presented, and in order to validate the behaviour patterns of these model variables, an example will now be introduced. This example is characterised by the following parameter values and by the market conditions shown in Figure 9.5, where three possible scenarios for demand and spot price are presented:

L	1 week	ss_{fc}	2 weeks
Lsc	1 week	ss_{fc}	2 weeks
Lfc	1.5 weeks	ss_{ufc}	2 weeks
$Lusc$	2 weeks	ss_{sc}	0 week
α	0.5 dimensionless	ss_{cp}	1 week
β_S	1 ½ days	Dsc	20 M units/quarter
Dfc_t	Within the range [1,5] M units/quarter		

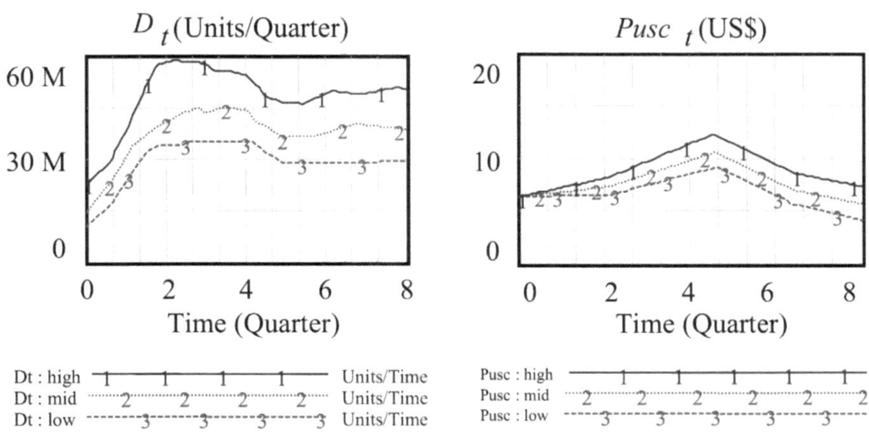

Figure 9.5. The scenarios for the critical part price and demand

Results for some operational variables of the model for this example are presented in Figure 9.6.

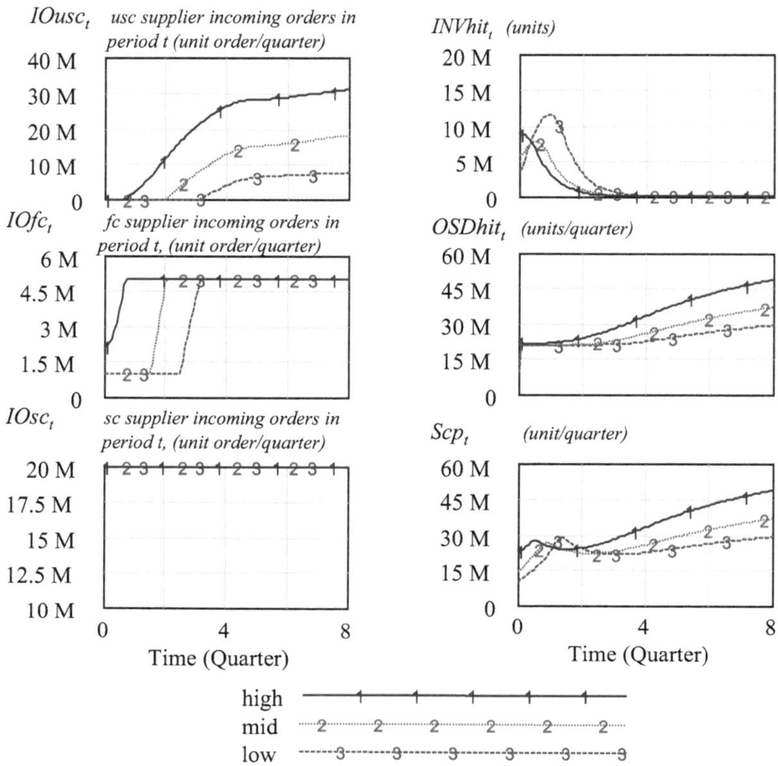

Figure 9.6. Materials and information flow variables

9.3 Modelling Accountability of the Procurement System

The accountability of the procurement system in Figure 9.4 has to be established based not only on how conveniently we purchase parts from suppliers but also taking into account other factors such as:

1. The average price (AP_t) of a unit delivered to the production lines from the buffer inventory *INVhit*.
2. The inventory holding cost of units in hubs, in transit inventories, or in the buffer inventory, over time (IHC_t). Note that the inventory holding cost will include staging, warehousing, flooring, losses, devaluation, and incentives paid to compensate network partners for value losses.
3. The chronological value of money (*rfir*).

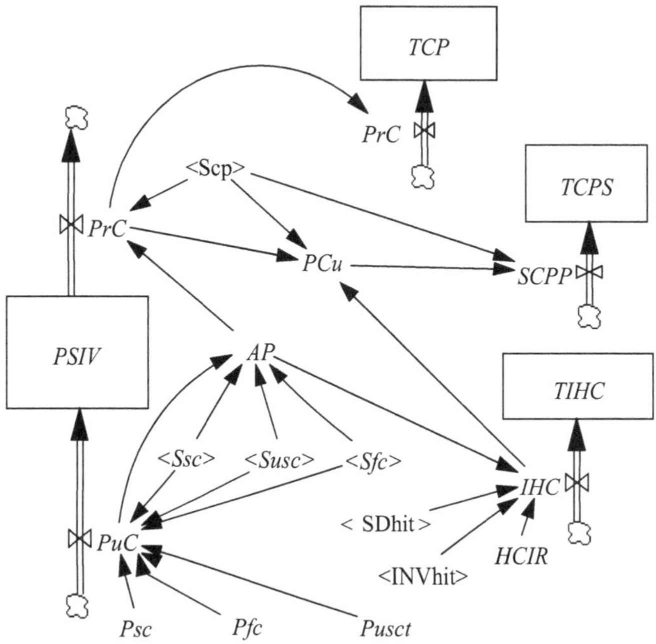

Figure 9.7. A stock and flow diagram for the financial variables of the procurement system model

As was done in the previous section, the notation and definition of the main financial variables and parameters of the model is now presented as follows

Nomenclature of financial variables:

$PSIV_t$ Procurement system inventory value in t
PuC_t Purchase cost from suppliers in period t
PrC_t Cost of parts delivered to the production line in period t
PCu_t Cost of the procurement system per unit delivered to the production line in period t
IHC_t Inventory holding cost in period t
$SCPP_t$ Procurement system cost in period t
AP_t Average parts purchase price in t
$TCPS_t$ Net present value of the total cost of the procurement system from 0 to t
TCP_t Net present value of the total cost of parts delivered to the production line from 0 to t
$TIHC_t$ Net present value of the total inventory holding cost from 0 to t

Parameters:

HCIR Inventory holding cost interest rate
rfir Risk free interest rate

According to the previous paragraph, and Figure 9.7, the following equations (Equations 9.19, 9.20 and 9.21) establish the monetary balance in the inventory of parts within our procurement system:

$$PSIV_t = PSIV_{t-1} + PuC_t - PrC_t \quad (9.19)$$

$$PuC_t = Ssc_t * Psc_t + Sfc_t * Pfc_t + Susc_t * Pusc_t \quad (9.20)$$

$$PrC_t = Scp_t * AP_t \quad (9.21)$$

In order to establish this balance, we had to estimate the average price of parts (Equation 9.22), to set up the cost of a part delivered to the production lines from the procurement system (assuming parts valuation to the current purchase cost):

$$AP_t = PuC_t / (Ssc_t + Sfc_t + Susc_t) \quad (9.22)$$

For a period of time, the cost of the procurement system (Equation 9.23) will be the sum of the cost of parts delivered to the production lines plus the inventory holding cost:

$$SCPP_t = PCu_t Scp_t \quad (9.23)$$

$$PCu_t = (IHC_t + PrC_t) / Scp_t \quad (9.24)$$

$$IHC_t = (SDhit_t + INVhit_t) AP_t HCIR_t \quad (9.25)$$

Once the previous variables are obtained, the net present values of the streams we are concerned about can be calculated:

$$TCP_t = NPV(PrCt, rfir, t, 0) \quad (9.26)$$

$$TIHCt = NPV(IHCt, rfir, t, 0) \quad (9.27)$$

$$TCPS_t = NPV(SCPPt, rfir, t, 0) \quad (9.28)$$

where NPV(A_t,B,t,0) used in Equations 9.26, 9.27 and 9.28 is a function returning the net present value of the stream A_t, from time 0 to *t*, obtained at time 0, and computed using discount rate *B*.

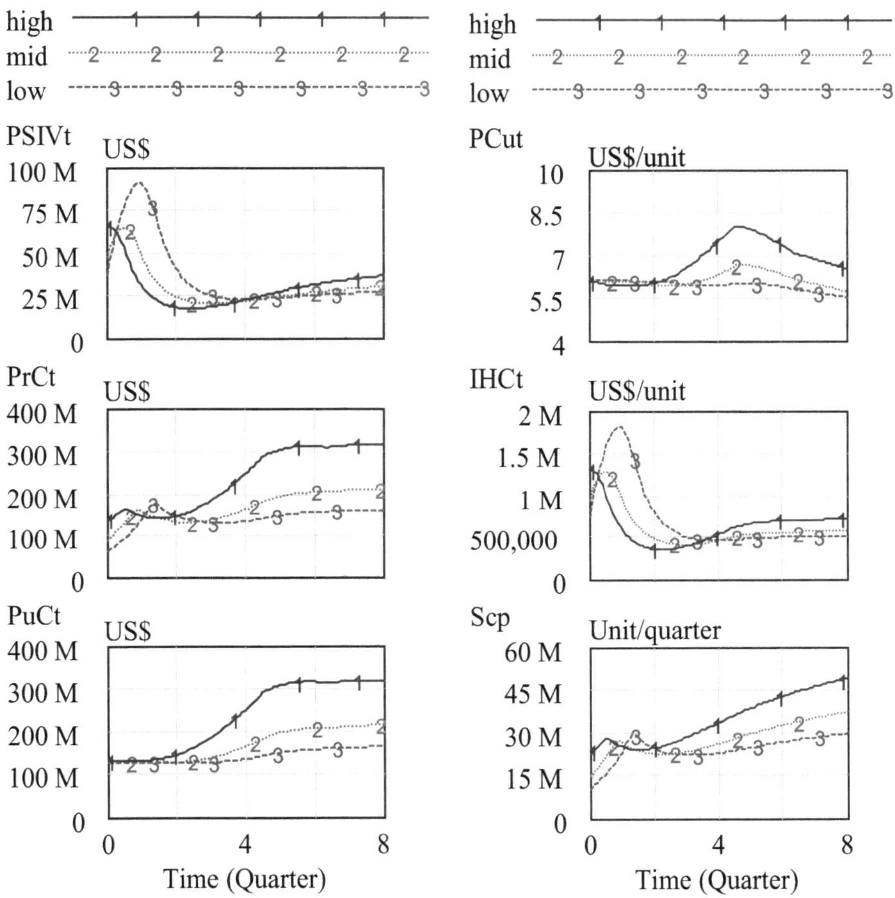

Figure 9.8. A stock and flow diagram for the financial variables of the procurement system model

In Figure 9.8, we present some of these variables, for the example in Section 9.2.3.

9.4 Modelling Forward Contract with Suppliers

In the case of price increase scenarios, such as those presented for trailing edge technology parts, a firm commitment to buy will normally provide considerable return in supplier pricing, compared to the prices that the buyer would experience over the contract period. In the absence of contractual commitment to purchase on the part of the buyer, and a corresponding dedication of capacity commitment from the seller, a low negotiated price motivates suppliers to shift capacity and parts allocation to higher-margin customers when world demand exceeds world supply.

Contract terms must be negotiated to reward suppliers who honour delivery commitments in a situation of decreasing capacity, fewer suppliers and highly competitive markets. As previously considered, we shall assume that for fixed flow, structured (S) contracts with the suppliers, the forecast for the year is fixed and linearised (these conditions provide the best possible price from suppliers, since they are relieved of the costs of demand volatility in their inventory levels and capacity utilisation through the contract horizon; the terms could further discount the unit price for early payment).

The overriding purpose of structured contracts is of course to secure capacity and protect profit margins in the face of almost certain inability of world capacity to meet expected demand, and corresponding increasing prices. The supplier who negotiates a flexible flow (F) type of contract experiences volatility in demand, and not only as a consequence of the final product market volatility, but also as a consequence of the supply chain structure and corresponding bullwhip associated to it.

In order to structure the contracts with the suppliers (S and F), a valuation procedure is to establish the deal as a series of forward contracts for each delivery period (roughly speaking, a forward contract is a contract to buy or sell at a price that stays fixed for the life of the contract), and then use the conventional valuation approaches for their financial assessment (the value of a forward contract is the difference between the futures price and the forward contract price, discounted to the present at the short-term interest rate) [7]. By doing so, current price will be used as a basis (BP, in US$) for the valuation since the product is considered to be purchased at present (we would use current price for a quantity of parts corresponding to the total annual purchase), discounted based on expected (forecast) delivery, but deferring payments until the time of delivery although the contract would be binding.

The real purchase cost of a unit of critical part will take into account the cost of borrowing (as "cost of borrowing", the cost of capital could also be used) the money (*r, in %*) until the part is delivered by the supplier, plus the usual net payment terms the procurement system has with its suppliers (*PT, in weeks*). If we consider that all deliveries for the year were paid upfront, borrowing cost (notice how the borrowing cost for a year is also the real value of having the supplier to hold inventory or capacity during that period) should be calculated for each part delivery period (*t, in weeks*).

Delivery price for a part delivered in period t under a flexible contract (*DPfc*) would then be calculated, assuming continuous compounding, according to Equation 9.29, which is replacing Equation 9.1, after articulating the forward contract:

$$DPfc_t = f(t, BP, PT, r) = BPe^{r(t+PT)} \qquad (9.29)$$

Note how, by paying the marginal amount $BP(e^{r(t+PT)}-1)$, we limit the cash investment and price risk in the purchase of a part in period t. In this case, Equation 9.20 would be transformed into Equation 9.30:

$$PuC_t=Ssc_t{*}Psc_t+Sfc_t{*}DPfc_t+Susc_t{*}Pnsc_t \tag{9.30}$$

and this will also change totally the valuation of inventories within the system, and the total cost of procurement.

9.5 Modelling Commodity Options Contracts with Suppliers

A commodity option is an option to buy (call option) a fixed quantity of a specified commodity at a fixed time in the future and at a fixed price. It differs from a security option in that it can't be exercised before the fixed future date. This is an "European Option" rather than an "American Option" [7]. A commodity option differs from a forward contract because the holder of the option can choose whether or not he wants to buy the commodity at the specified price. Note that with the forward contract he has no choice: he must buy it, even if the spot price at the time of the transaction is lower than the price he pays. There are five factors affecting the value of a call option [8]. Its value increases with:

- an increase in the price of the underlying commodity;
- the exercise price (the lower it is the more valuable the option will be);
- the time to maturity (with more time to maturity the greater the chance that the maturity price will be higher above the exercise price),
- the variance of the underlying commodity price (the greater the variance the greater the probability that the commodity price will exceed the exercise price on the upside, while on the downside the minimum option value is zero); and
- the risk free rate (if the risk-free rate increases and nothing else changes, then a call must be worth more because the discounted present value of the exercise price declines).

Now it is assumed that an option contract for a commodity is structured, this commodity is to be delivered in the period t, where the exercise (strike) price of the option is the delivery price of the commodity in that period $DPfc_t$ (estimated in Section 9.2). Once a forward contract is a contract to buy at a price that stays fixed for the life of the contract, assume that a fraction c of the current price BP is offered as an option price to buy at $DPfc_t$ in period t. In order to articulate a simple proposal to suppliers, c will be considered constant for the whole contract review period.

In this case, Equation 9.31 would replace Equation 9.20, and again the valuation of inventories within the system would totally change, as well as the total cost of procurement:

$$PuC_t=\begin{cases} Ssc_t{*}Psc_t+Sfc_t{*}((1+c)DPfc_t)+Susc_t{*}Pusc_t, & DPfc_t\leq Pusc_t \\ \\ Ssc_t{*}Psc_t+Sfc_t{*}(\,cDPfc_t+Pusc_t)+Susc_t{*}Pusc_t\ , & DPfc_t> Pusc_t \end{cases} \tag{9.31}$$

Note that now there would be no floor to buy parts from the flexible supplier, while the cap would now be conditioned by the number of options written against the commodity in the contract for each period *t*, estimated according to the forecasted purchase for period *t* (Sfc_t) from the flexible supplier.

Equation 9.31 shows how in case of price decrease below the spot price, no option would be executed and therefore price of the parts delivered would be the spot price at that time, assumed to be purchased from the same flexible supplier. The option approach reserves capacity, but does not legally bind to take inventory. It limits risk, but requires upfront payment.

Note that both parties may guarantee their ability to fulfil the contract. The supplier may realise that although his capacity may not be utilised if there is a demand shortage, he needs to have the capacity available at maturity, just in case the option is executed. This could be an issue to discuss in the contract.

9.6 Selecting a Suitable Contract Portfolio with Suppliers

In this section the previous simulation model is used to help in the study and selection of a suitable portfolio of contracts with suppliers. A contract proposal from a flexible supplier will be analysed.

The idea is to use the model to compare this proposal against a forward contract, and against an option contract that could be alternatively structured with the supplier. In the latter case, it is required to value the price of the option that would be reasonable to pay, according to Section 9.5.

The portfolios to analyse are summarised in Table 9.1, and mainly portfolios B, C and D will be compared assuming a strategic decision to go for the three tier sourcing model.

In Figure 9.9, the supplier proposal for the price of parts delivered under a flexible supply contract for a review period of two years is presented. Conditions for parts delivery in this example include a floor of 750,000 units/week, and a cap of 1.6 M units/week. In the same figure the forecast for the spot price *Pusc* (in this example, assuming 20% spot price volatility around the forecast) is shown, together with the model estimation for the forward contract delivery price, and the model estimation of the price of a part in case that an option contract is written against it (note that the price paid for the call option is included in *Pfc[option contract]*).

For this example, it will be considered that the inventory holding cost rate will be 2% per week, the risk free interest rate 8% per year, and that the payment terms of the contract would include a 4 weeks of sales outstanding with the supplier. Conditions for total demand are the same as in Figure 9.5 for the “mid” case scenario, but we have now transformed our time unit to weeks.

Table 9.1. Contracts portfolios for simulation analysis

	Portfolio A	Portfolio B	Portfolio C	Portfolio D
Short term contract. (ST contracts)	For the whole forecast	For the remaining amount to cover the forecast (once S and F contracts are articulated)	For the remaining amount to cover the forecast (once S and F contracts are articulated)	For the remaining amount to cover the forecast (once S and F contracts are articulated).
Flexible contract. (F contracts)	No	Based on a proposal (price, quantity) from the supplier	Based on a forward contract structured with the flexible supplier	Based on an options contract structured with the flexible supplier
Structured contracts (S contracts)	No	Constant delivery based on minimum demand forecasted	Constant delivery based on minimum demand forecasted	Constant delivery based on minimum demand forecasted

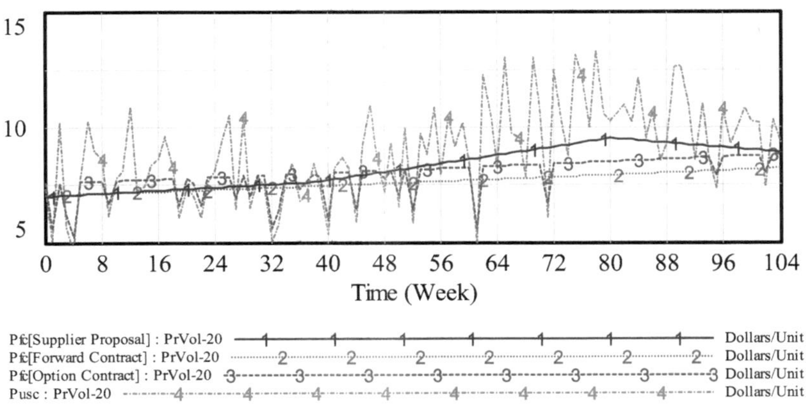

Figure 9.9. Price offered by the suppliers compared to forward contract and spot prices

Numerical optimisation techniques are now applied with the simulation model, in order to find out the value of *c* (fraction of the current price *BP* to pay for an option to buy a part at *DPfc* in period *t*), with the condition that the portfolio D will have the same value as portfolio C, now assuming no price volatility.

Once again, regarding numerical optimisation techniques, the direct-search method which does not need to evaluate the gradient is most suitable for the analysis of dynamics of complex nonlinear control systems such as the one we are dealing with. In this section we have used the Powell method [9], well known to have an ultimate fast convergence among direct-search methods. The basic idea behind Powell's method is to break the N dimensional minimisation down into N separate 1D minimisation problems. Then, for each 1D problem a binary search is implemented to find the local minimum within a given range. Furthermore, on subsequent iterations an estimate is made of the best directions to use for the 1D searches. It is proved that, with the Powell method, at most m iterations, where m is the number of parameters to be estimated, yield the optimal solution to the problem with cost function of quadratic form if the directions of m-dimensional vectors are linearly independent at every iteration step.

The NPV of the total savings over time of the portfolio D, compared with the portfolios C and B (calculated as follows: TCP_t[Portfolio C]–TCP_t[Potfolio D], and TCP_t[Portfolio B]–TCP_t[Potfolio D], respectively) are presented in Figure 9.10.

The NPV of the total savings in cost of parts of the portfolio D, compared with the portfolios C and B (calculated as follows: $TCPS_t$ [Portfolio C]–$TCPS_t$ [Portfolio D], and $TCPS_t$ [Portfolio B]–$TCPS_t$ [Portfolio D], respectively) are presented in Figure 9.11.

The NPV of the total savings in inventory holding cost of the portfolio D, compared with the portfolios C and B (calculated as follows: $TIHC_t$ [Portfolio C]–$TIHC_t$ [Portfolio D], and $TIHC_t$ [Portfolio B]-$TIHC_t$ [Portfolio D], respectively) are presented in Figure 9.12.

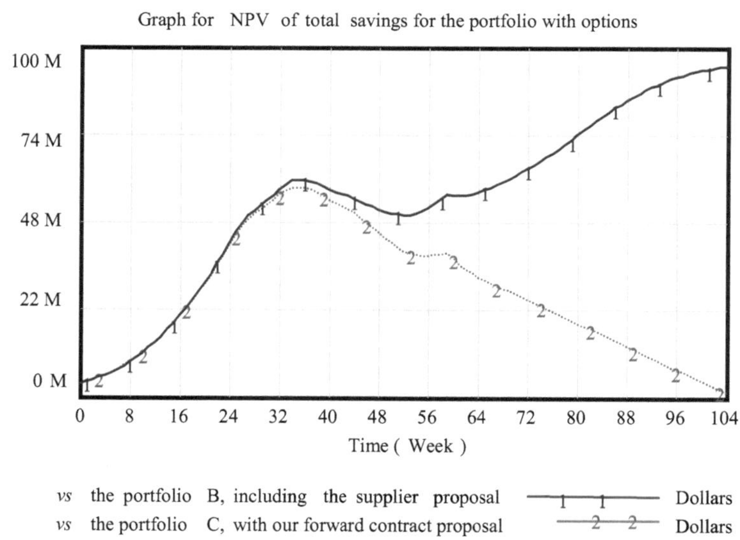

Figure 9.10. Total savings for the portfolio D with the option price of 7.3% of the strike price (delivery price) in US$

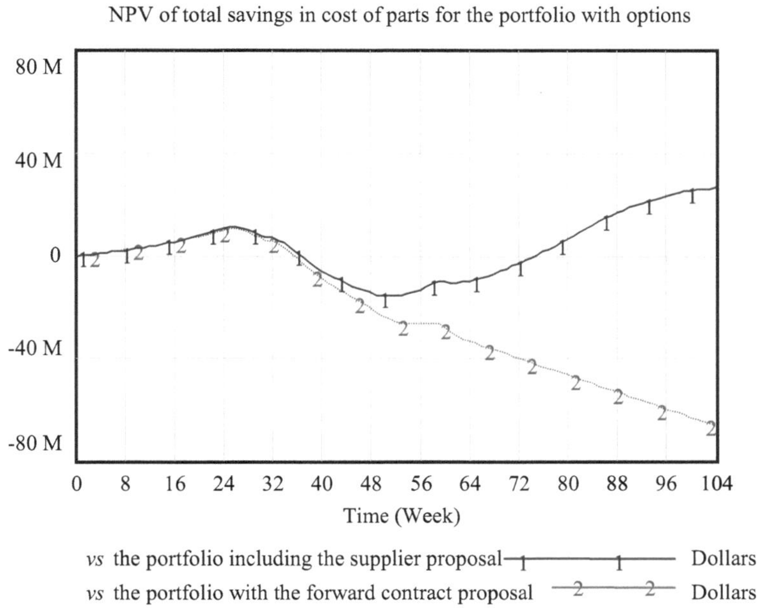

Figure 9.11. Savings in cost of parts for the portfolio D, and option price of 7.3% of the strike price (Delivery price) in US$

Discussing previous results it can be seen how, assuming no volatility in price forecast for the simulation horizon, paying 7.3% of the delivery price of the parts (price of parts for the forward contract) for an option to buy the parts in time t, at that price, would produce more or less the same savings over the 2 years, compared to structuring a forward contract with the flexible suppliers (Figure 9.10).

Figure 9.11 shows how the portfolio including the option contract with the flexible suppliers would have more cost of parts than the forward contract at the end of the simulation, but better cost than the supplier proposal over the 104 weeks.

In Figure 9.12, however, we see how the savings in inventory holding cost of the portfolio D, *vs* the portfolio C, compensate the difference regarding the cost of parts in Figure 9.11. In Figure 9.12, we see how portfolio D is also more beneficial than portfolio B in terms of inventory holding cost.

As mentioned in Section 9.5, we may expect that for an increasing price volatility, the portfolio including the options contract may provide better results. Figure 9.13 presents the savings of the portfolio D, *vs* the portfolio C (TCP_t[Portfolio C]–TCP_t[Portfolio D]) in case of volatility in price increases. Results show how for price volatility over 20%, benefits of the options contract are very significant.

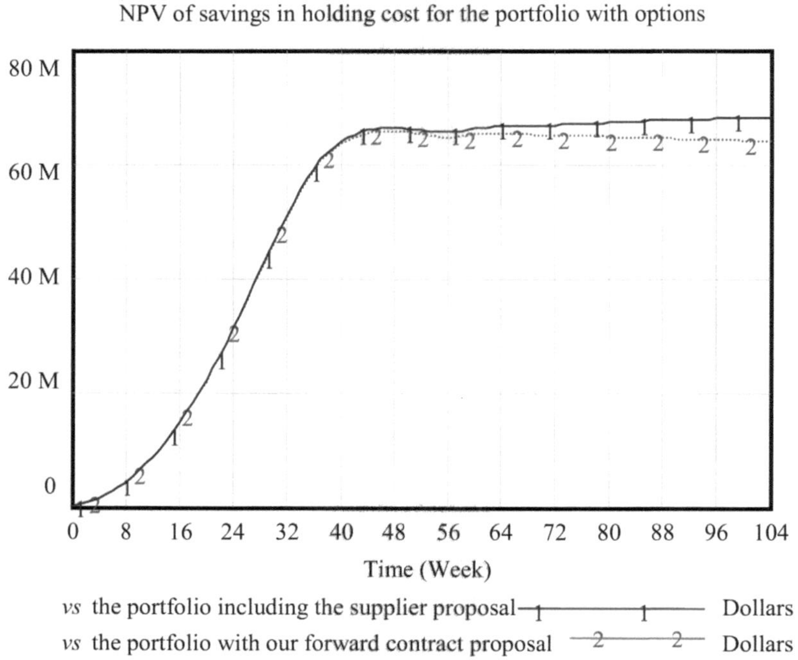

Figure 9.12. Savings in inventory holding cost for the portfolio D, and option price of 7.3% of the strike price (Delivery price) in US$

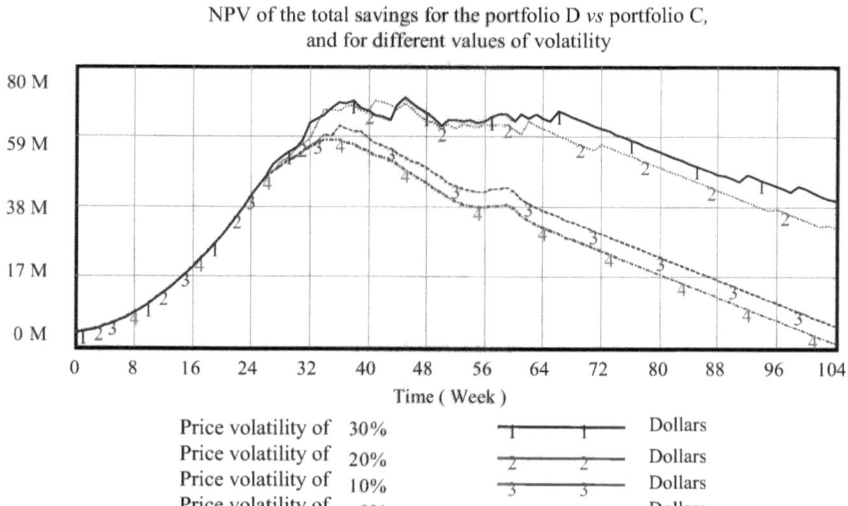

Figure 9.13. Incidence of price volatility in the value of the portfolio including the option contracts (D) with the flexible suppliers, compared with the portfolio including the forward contract (C)

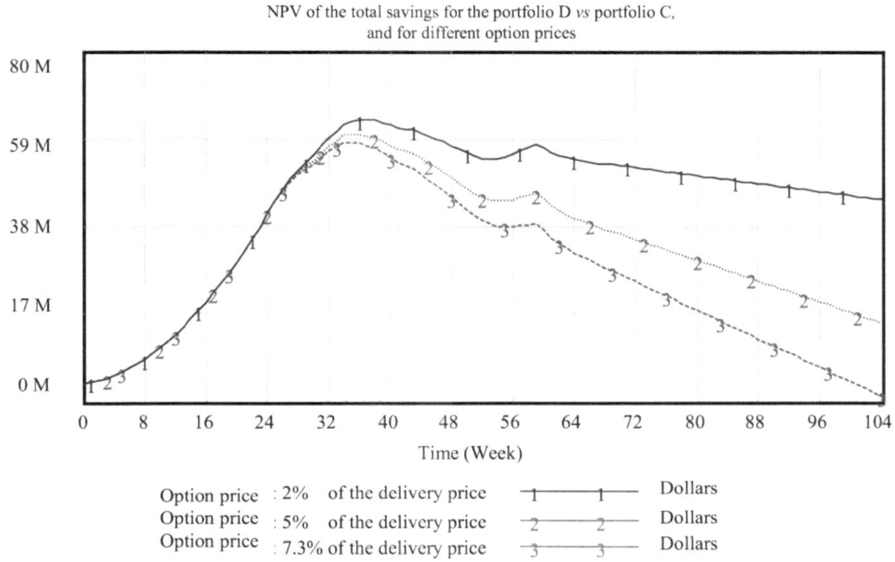

Figure 9.14. Incidence of price volatility in the value of the portfolio including the option contracts with suppliers, compared with the portfolio including the forward contract

Figure 9.14 shows how the net present value of the savings for the portfolio with the option contracts *vs* the one with the forward contract (TCPt [Portfolio C]–TCPt [Potfolio D]), as a function of the price paid for the option.

Figure 9.14 could also help to articulate a contract with the supplier. For example, paying 2% of the delivery price, for the option to buy at that price, could be beneficial for the whole contract horizon even if we pay up to US$38M approximately upfront to the flexible supplier (for the contract review period of 2 years).

9.7 Managerial Implications of the Work

Managerial implications of this work are related to the design of better policies in order to maintain suitable relationships with suppliers. Each type of relationship will pursue a different objective, and will have some different tradeoff implications that we have to acknowledge. All these ideas have been summarised in Table 9.2, where, at the same time, the consideration of certain types of policy and managerial practices for each type of supplier is recommended.

The tiered approach allows, besides protecting from risk, the selection and implementation of convenient deals with each specific supplier, protecting the supplier if needed, aligning him to serve specific market needs, or market strategies.

Table 9.2. Different policies and practices considerations with the suppliers

Tier	Objective	Trade-off	Policy and practices considerations
Tier 1, Structured contract, fixed flow	Lowest possible price, highest capacity utilisation	Commitment to order at pre-specified rates throughout the contract period	• Openness with supplier about the savings from fixed cost recovery when production is kept linear • Agreement to own and manage the excess inventory • Relaxation of lead times, as along as delivery is on time • For suppliers of trailing edge technology, encourages them to continue to maintain capacity, and encourages mutual discussion with their buyers about the technology end-of-life management
Tier 2, Flexible contract, variable flow	Limited exposure to price increases, payment for access to additional extra capacity	Allows for order and delivery adjustments when actual demand does not match forecasts	• Acknowledging the risk from volatile markets, especially focused on securing capacity to avoid stockouts and missed shipments if demand exceeds forecasts • Agreement on lead times when forecasts and order levels change • Negotiation about the location, ownership, and management of safety stock • Negotiation to limit the timing and range of price changes, understanding the impact on inventory valuation in various scenarios • Negotiation to limit the timing and range of forecast changes, understanding the impact on safety stock policy and storage location
Tier 3, Unstructu-red, open contract, demand-driven flow	Flexibility to buy as needed, where capacity or excess parts are available	No assurance on capacity or parts, may have to pay full spot price	• Suppliers pre-qualified and procurement processes in place • Lead times pre-defined, with a good expectation of acceptable service levels and quality • Membership in internal or external trading consortiums, or in commercial trading hubs, to quickly meet demand when there is a worldwide part shortage and capacity • Ability to select the source by comparing lead time and price alternatives in terms of their impact on overall inventory levels, values, service levels, and profit

Specifically, the model that is presented in this chapter can simulate the tradeoffs available to managers in the various contract structures. In this sense, it is very helpful to understand the implications of the different contract parameters, when markets conditions may change, and for metrics selected by the decision maker. This model is also suitable to simulate the combined impact of a portfolio as a whole, in the context of the overall supplier relationship.

9.8 Concluding Remarks of the Chapter

This chapter presents a simulation model to assess and compare portfolios of contracts with suppliers within a multi-tier sourcing framework for the management of strategic parts. The implications of setting different levels of parts inflow from each type of supplier are studied in terms of inventory holding cost and parts cost of purchase. The model shows the implications of price volatility and call option valuation, in the global assessment of a portfolio. This model and framework can be used for the proactive design of the contracts with suppliers of strategic parts, and for the analysis of the problem from both the supplier and buyer perspectives. A full valuation of inventory and related carrying costs, inclusion of accepted predictive statistical tools, and tracking the cost of capital in valuing cash flows over time, all allow this model to support fully the creation and management of options contracts in lieu of forward contract structures for flexible demand management.

9.9 References

[1] Crespo Marquez A, Blanchar C, (2004) The procurement of strategic parts. Analysis of a portfolio of contracts with suppliers using a System Dynamics simulation model. International Journal of Production Economics, 88: 29–49.

[2] Chen F, Drezner Z, Ryan JK, Simchy-Levy D, (1998) The bullwhip effect: Managerial insights on the impact of forecasting and information on variability in a supply chain. In: Tayur S, Ganeshan R, Magazine M, (Eds.) Quantitative Models for Supply Chain Management, International Series in Operations Research and Management Science, 17: 417-439. Boston: Kluwer Academic Publishers.

[3] Sanders NR, 1994. Forecasting practices in United-States corporations. Survey results. Interfaces, 24 (2): 92–100.

[4] Makridakis S, Wheelwright S, Hyndman R, (1998) Forecasting Methods and Applications. New York: Wiley.

[5] Tversky A, Kahneman D, (1974) Judgment under uncertainty. Heuristics and biases. Science, 185: 1124–1131.

[6] Sterman JD, (1989) Modelling managerial behaviour: Misperceptions of feedback in a dynamic decision making experiment. Management Science, 35(3): 321–339.

[7] Black FT, (1976) The pricing of commodity contracts. Journal of Financial Economics, 3: 167–179.

[8] Weston JF, Copeland TE, (1989) Managerial Finance. 8th. Edition. Chicago: The Dryden Press.

[9] Powell MJD, (1964) An efficient method for finding the minimum of a function of several variables without calculating derivatives. Computer Journal, 7(2): 155–162.

10

Modelling Back-end Issues in Manufacturing

10.1 Introduction to the Modelling of Manufacturing Issues

Simulation has been frequently used in manufacturing because it allows alternative designs and control policies to be tried out on the model during the preparatory phase of the physical plant. It helps to reduce cost and risk of large scale errors. Simulation approaches are also used during the operational phase of the manufacturing plants to find better ways to operate, and these studies may be one point in time exercises or may be part of a periodic check on the running of the system [1].

At present, mass customisation has placed very high pressure on manufacturers. Manufacturing systems need to deliver high-volume and high-quality goods at very low cost to meet customers needs. As a consequence of this, large investments in production plants and equipment were accomplished and there is a clear need to ensure that these production systems operate as expected. That is why computer simulation methods have found in manufacturing an important field to develop and to contribute to the design and optimisation of these systems.

There are many examples of the use of dynamic simulation across most manufacturing sectors including semiconductor wafer fabrication, beverages, pharmaceuticals, automobile manufacture, etc.

In this chapter, a semiconductor facility has been selected to illustrate the potential use of dynamic simulation techniques within this area. In the example presented in the next sections, dynamic simulation is compared to mathematical programming techniques as a proper alternative to solve a specific problem related to the maintenance scheduling of different toolsets. In a semiconductor wafer facility, tool availability, in turn, determines factory capacity and serves to drive factory performance in terms of outs, inventory, cycle time and work in process (WIP) velocity [2]. In order to ensure a certain system's dependability over time and at a certain cost, preventive maintenance plans need to be designed, scheduled and implemented.

This example has been selected among other work done in this field only because of personal interest for maintenance engineering and for the development of this area.

10.2 Case Study in Semiconductor Fabs

This case study reviews approaches for maintenance planning and scheduling in semiconductors fabrication facilities (fabs), and focuses exclusively on the scheduling problem. In this context, the complexity of the modern semiconductor manufacturing processes, as well as the need for realistic considerations when modelling their short term availability and reliability, render analytic methods very difficult to be used within these production environments. Simulation methods allow modelling the behaviour of these complex systems under realistic time-dependent operational conditions and may be very suitable tools with which to approach their short term preventive maintenance scheduling problem.

This work discusses the opportunity to use Monte Carlo continuous time simulation modelling to improve preventive maintenance scheduling in these environments. This technique allows the assessment of alternative scheduling policies that could be implemented dynamically on the shop floor. Policies considered will be based not only on current manufacturing tools status but also on several operating conditions of the wafers production flow.

The idea is to compare and discuss the benefits of the different scheduling policies using a simulation model and, in order to do so, measures of performance will be used, treating simulation results as a series of real experiments using statistical inference to reach reasonable confidence intervals.

10.3 Introduction to the Case Study

The design of a preventive maintenance plan will take into account primarily (see Figure 10.1):

1. the production plan,
2. the tool's failure dynamics,
3. the operating conditions of the process and
4. the different possible maintenance actions and their consequences according to required investments in instrumentation, diagnostic and repair tools, etc. [3].

When considering repairable systems and for finite time periods studies, Semi-Markovian Decision Processes (SMDP) have proven good capabilities to face this problem (see, for instance: [4–9]).

Semi-Markovian models offer a good trade-off between the complexity of the formulation (mathematical format and data requirements) and results provided in terms of detailed production systems behaviour replication. In cases where the maintenance planning design problem is modelled as a SMDP, it has also been shown [6, 10–12] (the reader is also referred to classical literature such as Bellman [13] and Howard [14]) that the utilisation of dynamic programming as the optimisation technique offers reasonable results in the calculation of the optimal preventive maintenance plan.

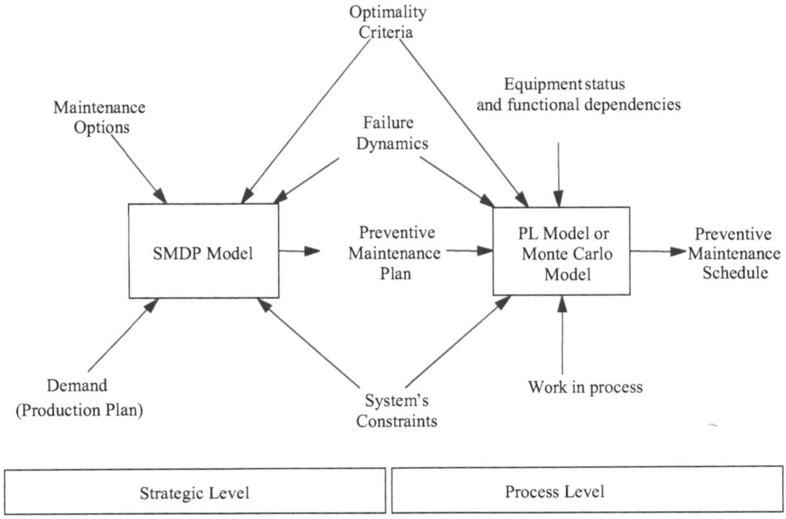

Figure 10.1. Preventive maintenance scheduling process

Once the optimal preventive maintenance plan is determined, the next problem is to schedule short term preventive maintenance activities. In order to deal with this problem the analyst must take primarily into account:

1. the preventive maintenance plan;
2. the status of the production system (i.e. state variables: wips);
3. the tool operating condition;
4. the possible functional dependencies among tool and tool components; and
5. the system failure dynamics.

For semiconductor wafer facilities, this problem has been already approached using linear programming models [15]. These LP models have a planning horizon shorter than the time between two equal maintenance activities of a manufacturing tool. Of course, it is assumed that the set of maintenance activities to carry out on a tool is known (maintenance plan) and the moment in time when each activity will be started is determined. This is done by minimising an objective function based on cost elements.

After discussing pros and cons of the previous LP models approach, this case study will focus on a second approach with which we propose to solve this problem; Monte Carlo (stochastic) dynamic simulation [11]. The idea behind this method is the generation of certain random and discrete events in a computer model in order to create a realistic timeframe scenario of the system. Therefore the simulation of the system's process will be carried out in the computer, and estimates will be made for the desired measures of performance [16]. The simulation will then be treated as a series of real experiments, and statistical inference will be used to estimate confidence intervals for the performance metrics.

The events can be simulated either with variable time increments (discrete event simulation) or with fixed time increments, at equidistant points of time (continuous time simulation), as discussed in the first part of this book.

In this case study, of course, continuous time simulation technique is used. This simulation will evaluate the system state every constant time interval (Δt); the new system state will be recorded and statistics collected. Then the time is incremented another Δt, and so on. The simulation software tool used was VENSIM simulation environment (Ventana® Systems [27]), which has special features to facilitate Monte Carlo type of simulation experiments, and to provide confidence interval estimations.

The Monte Carlo simulation method allow us [17,26] to consider various relevant aspects of systems operation which cannot be easily captured by analytical LP models such as K-out-of-N, redundancies, stand-by nodes, aging, deteriorating repairs, repair teams or component repair priorities.

Once the stochastic simulation model is built, there are multiple techniques to search for optimal solutions of the preventive maintenance scheduling problem. A basic classification of search techniques could divide them into:

- global techniques such as simulated annealing, or genetic algorithms; and
- local techniques such as hooke-jeeves search, nelder-mead simplex algorithm, or tabu search.

Concerning the criteria to search for optimal schedules, as well as system's availability, we will try to reduce the variability of the production flow, because it lowers the performance of the manufacturing system, increasing the cycle time and the work in process inventory levels [18]. These ideas have recently been applied by different researchers when dealing with the problem of finding optimal maintenance schedules (see, for instance: [19–23]).

The rest of the chapter is organized as follows: Section 10.4 discusses in detail the pros and cons of previous LP model approach to deal with the preventive maintenance scheduling problem in semiconductor fabs. Section 10.5 and 10.6 are dedicated to introduce and to formalise the continuous time Monte Carlo dynamic modelling of a semiconductor manufacturing tool set's preventive maintenance activities. Section 10.7 describes several configuration examples that will be used in the simulation study and Section 10.8 is given over to present and discuss results of the simulation study. Finally Section 10.9 is devoted to summarising the managerial implications of this work, interesting findings and some useful directions for future research.

10.4 Pros and Cons of LP Models to Deal with PM Scheduling

An approach to the low-level scheduling of preventive maintenance activities in semiconductor fabs proposed in the literature formulates the problem as a mixed integer program [15]. This approach allows for the consideration of resource constraints and work-in-process inventory, and takes as input the preventive maintenance windows determined in high-level planning models. In particular the mathematical model is formulated as follows:

Notation:

N	Number of periods in the planning horizon
M	Number of tools under consideration
ρ_i	Number of PM tasks or activities scheduled for tool *i* over the horizon
n_i^l	Maximum time at which PM task l on tool *i* can be started to satisfy its time window
m_i^l	Minimum time at which PM task l on tool *i* can be started to satisfy its time window
a_i^l (t)	Binary decision variable that takes the value 1 if PM task l is started on tool *i* at time *t*, and 0 otherwise
$\underline{a}_i(t)$	Action vector for all tasks on tool *i*
$\underline{a}(t)$	Action vector for all tasks on all tools
$\underline{\beta}_i(t)$	Accompanying status vector for PM task vector $\underline{a}_i(t)$. They are used specifically for those PM tasks whose duration is 2 days; specifically, we have $\beta_i^l(t+1) = a_i^l(t)$ for all such tasks
$\underline{\beta}(t)$	Accompanying vector for all tasks on all tools
$V_i(t)$	Availability of tool *i* at time *t*
K_i	Wafer throughput coefficient for availability of cluster tool *i*
$I_i(t)$	Workload level (total in queue and in process) for tool *i* at time *t*
$d_i(t)$	Workload from upstream operations for tool *i* at time *t*
b_i	Weighted profit coefficient for availability of cluster tool *i*
C_i^I	Cost coefficient for inventory on tool *i*
C_i^l	Cost of performing PM task *l* on tool *i*
L_i	Workload upper limit for tool *i* at any time
R(t)	Total resources available for PM at time *t*
$f_i(\underline{a}_i(t), \underline{\beta}_i(t))$	Availability function for tool *i* under PM status vector $\underline{a}_i(t)$ and $\underline{\beta}_i(t)$
$r(\underline{a}(t), \underline{\beta}(t))$	Resource function, computing resource required by tasks status vector $\underline{a}(t)$ and $\underline{\beta}(t)$

Mixed-Integer Programming Formulation

$$\max_{a(t)} \sum_{t=1}^{N}\sum_{i=1}^{M}\left(b_i V_i(t) - C_i^I I_i(t) - \sum_{l=1}^{\rho_i} C_i^l a_i^l(t) \right)$$

subject to:

(1) $\sum_{t=1}^{n_i^l} a_i^l(t) = 1$ for all PM activity l on tool i

(2) $\sum_{t=1}^{m_i^l} a_i^l(t) = 0$ for all PM activity l on tool i

(3) $V_i(t) = f(a_i(t), \underline{\beta}_i(t))$ for all tool i and period t

(4) $I_i(t+1) = I_i(t) - X_i(t) + d_i(t)$ for $i = 1,2,...M$, and $t = 1,2,...,N$

(5) $X_i(t) \le K_i V_i(t)$ for all $i = 1,2,...M$, and $t = 1,2,...,N$

(6) $I_i(t) \le L_i$ for all $i = 1,2,...M$, and $t = 1,2,...,N$

(7) $r_k(t) = g(a_i(t), \underline{\beta}_i(t))$ for all resource k and period t

(8) $r_k(t) \le R_k(t)$ for all resource k and period t

(9) $V_i(t) \ge 0, I_i(t) \ge 0, X_i(t) \ge 0, r_i^k(t) \ge 0$, for all resource k, tool i and period t

(10) $a_i^l(t) \in \{0,1\}$ for all tool i, PM activity l and period t

Examining this formulation, it is appreciated that:

- The workload from upstream operations for tool i at time $t, d_i(t)$, is not a variable but an input parameter. Therefore, the different tools are linked by the resource constraints, but their operational interrelationship is not well captured by this model. In practice, the workloads from upstream operations for tool i at time t will depend on the PM actions taken in the various upstream tools in the process.
- There is no constraint in the formulation to represent the material flows from one tool to the next in the process. As a result, although inventories are measured and valued in the objective function, they are not accurately estimated.
- The IP model does not capture the dynamics of the manufacturing process and how PM decisions for different tools affect inventories and the availability of material downstream.
- It is claimed that "the objective of the LP model is to maximise the total availability (throughput) of cluster tools subject to the requirements that each PM activity (task) has to be performed within their specified time windows, and resource constraints have to be satisfied. For constraint tools, the objective is simply to maximise availability, not including inventory cost from WIP queued at the tools, while for non-constraint tools, the objective would be to match the availability with the demand pattern, in order to try to reduce inventory cost." This objective must then be achieved,

by appropriately choosing the objective coefficients, b_i and C_i^I. It can be argued, however, that the overall objective should be to maximise the throughput of the wafer manufacturing process and not the availability of individual tools. The constraining or bottleneck operations will change over time depending on preventive and corrective maintenance operations carried out or other quality problems. This makes it very difficult to estimate adequately the parameters b_i and C_i^I.

- Another major flaw of the approach is that uncertainty in yield and machine breakdowns is only captured through a constant K_i that translates the availability of tool *i* into its wafer throughput. (Although the authors do not specifically mention how this constant is determined, we interpret that it must account for yield loss due to uncertainty.) In any case, there is no direct consideration of uncertainty in the system.

Besides the previous points, it is important to emphasise that simplicity is the key for the successful implementation of a PM scheduling tool on the shop floor. The IP tool will provide maintenance technicians with a list of tasks to perform over the planning horizon. The plan is static, rigid, over that horizon and should be executed as given. The shop floor, however, is highly dynamic and the status of tools is continuously changing, sometimes predictably but also unpredictably.

A set of simple rules to guide operators' decisions in this environment can be more valuable than a fixed set of times and tasks to perform that do not account for the changing manufacturing environment. What is the technician to do if tool *i* is down for corrective maintenance at time *t* but scheduled for PM task 1 at time t+1? Common sense would make sure that the PM task is performed at the same time as the corrective one. The schedule of precise timings over the planning horizon would soon be disrupted.

10.5 Dynamic Simulation to Deal with PM Scheduling in Fabs

10.5.1 Introduction and Notation

A generic continuous time stochastic model for a tool's maintenance will now be built (to see a similar application of this modelling approach the reader is referred to Crespo Marquez *et al.* [24]). The notation will be as follows (note that this variable list could be later subscripted according to the number of tools in a tool set):

Tool status information related variables:

CA_t	Decrease in tool's age due to corrective maintenance action in *t*
LC_t	Time when the last corrective maintenance, for a tool in *t*, started
LP_t	Time when the last preventive maintenance, for a tool in *t*, started
PA_t	Decrease in tool's age due to preventive maintenance action in *t*
$\mathrm{RN_t}$	Random number within the interval (0,1), generated in *t*

T_t	Tool's age in *t*
TI_t	Increase of system's age in period *t*
TO_t	Decrease of system's age in period *t*
$\lambda(T_t)$	Failure rate of the system in *t*
A_t	Tool availability (1 available, 0 unavailable) at *t*
AA_t	All tools available (1 yes, 0 no) at *t*
SM_t	Scheduled maintenance (1 yes, 0 no) in period *t*
MB_t	Maintenance backlogged (1 yes, 0 no) at *t*
RM_t	Maintenance released (1 yes, 0 no) in period *t*
IFS_t	In-front stock status in *t*
FS_t	Finished stock status in *t*
$PTB_{t,i}$	Time that a PM action is being in backlog in *t*
$TBI_{t,i}$	Time increase (1 yes, 0 no) of a PM action in backlog in period *t*
$TBD_{t,i}$	Time increase (1 yes, 0 no) of a PM action in backlog in period *t*

Model parameters:

CT	Average time of a corrective maintenance action
n	Minimum age of the tool to do preventive maintenance actions
N	Maximum age of the tool to do preventive maintenance actions
PT	Average time of a preventive maintenance action
T1	Maximum time the tool operates without a failure
MxWi	Maximum desired in-front stock level to release maintenance

10.5.2 Modelling Tool's Age

The process first requires one to model the age of the system (*Tt*):

$$Tt = Tt\text{-}1 + TIt - TOt \tag{10.1}$$

We will consider that age will increase when the tool is available, i.e. we assume that available means "running", neither idling nor standing-by; therefore:

$$TIt = At \tag{10.2}$$

and age will decrease when the system is maintained:

$$TOt = \begin{cases} PAt, & \text{if } PAt <> 0 \text{ and } CAt <> 0 \\ PAt + CAt, & \text{Otherwise} \end{cases} \tag{10.3}$$

$$CAt = \begin{cases} Tt, & \text{if } \lambda(Tt) \geq RNt \\ 0, & \text{Otherwise} \end{cases} \tag{10.4}$$

where RN_t is a random number generated for every t within the range (0,1), $\lambda(T_t)$ is the failure rate of the system, and CA_t and PA_t are decreases in the system's age as a consequence of the corrective and preventive maintenance actions respectively.

10.5.3 Modelling Tool Availability

The conditions of a tool that will make it unavailable will be the corrective or preventive maintenance that is being carried out

$$At = \begin{cases} 1-(Pulse(LC_t,CT,t)+Pulse(LP_t,PT,t)), & if\ LC_t>0\ or\ LP_t>0 \\ 1, & Otherwise \end{cases} \quad (10.5)$$

Note that when $t=0$, $LCt = LPt=0$ (LCt and LPt are the times when the last corrective preventive maintenance started for a tool in t, respectively). The function Pulse is defined as follows:

$$Pulse\ (a,b,t) = \begin{cases} 1, & a<t<a+b \\ 0, & Otherwise \end{cases} \quad (10.6)$$

For instance, imagine that we have a tool set with two tools ($i=1,2$), and both of them need to be in operating conditions in order to maintain preventively one of them; then the condition to fulfil would be AA_t=1, where AA_t is defined as follows:

$$AA_t = \prod_{i=1}^{i=2} A_{t,\,i} \qquad with\ i=1,2 \quad (10.7)$$

10.5.4 Modelling Maintenance Activities Backlog

Although a tool's age or a wafer's yield may indicate starting a preventive maintenance action, it may be beneficial to leave the tool functioning while another tool is down (under corrective or preventive maintenance). Therefore, it is necessary to model the possible backlog of maintenance activities, *i.e.* activities which are due and waiting to be carried out by the maintenance department. This concept of maintenance backlog will be very practical when considering a feasible time range for PM actions to be carried out on a tool. We model the backlogged activities and the time they are backlogged as follows:

$$SM_{t,i} = \begin{cases} 1, & t_i/n = Int(t_i/n) \text{ and } t_i > 0 \\ 0, & Otherwise \end{cases} \tag{10.8}$$

$$MB_{t,i} = MB_{t-1,i} + SM_{t,i} - RM_{t,i} \tag{10.9}$$

In our model, maintenance actions will be scheduled according to Equation 10.8 when $T_t \geq n$. The backlog of maintenance activities is formalised as in Equation 10.9, where $RM_{t,i}$ represents the PM activity released (its value is 1 when the activity is released). A PM activity will only be released when a certain condition which characterises the preventive maintenance policy (defined later in this section) is met. The time a PM scheduled action is backlogged is being modelled to use it as a control variable and also to use it later as a performance measure. It is defined is as follows:

$$PTB_{t,i} = PTB_{t-1,i} + TBI_{t,i} - TBD_{t,i} \tag{10.10}$$

$$TBI_{t,i} = \begin{cases} 1, & RM_{t,i} = 0 \text{ and } MB_{t,i} > 0 \\ 0, & Otherwise \end{cases} \tag{10.11}$$

$$TBD_{t,i} = PTB_{t,i} \; RM_{t,i} \tag{10.12}$$

where $TBI_{t,i}$ and $TBD_{t,i}$ are increases and decreases in the time for which PM activity is backlogged, respectively. Equation 10.11 expresses how backlog time increases when the activity is not released while Equation 10.12 formalises decreases in time when PM is released. Therefore, this will later set the backlogged time to 0 in Equation 10.10.

10.6 Modelling Preventive Maintenance Policies

10.6.1 Overview

In this section, we model the way PM actions are released. The following policy options are considered:

- age based maintenance;
- age and availability based maintenance; and
- age and in-front buffer maintenance.

When a preventive activity is released, we will record this time (in LP_t) to allow downtime modelling as explained previously in Equation 10.5. Note that, in this example, we constrain the time that a backlogged activity can be backlogged so it will be released before a new preventive maintenance is scheduled. Also, regardless of the PM policy, Equations 10.13 and 10.1 will later set up tool's age to zero:

$$PA_{t,i}=T_t\, RM_{t,i} \tag{10.13}$$

10.6.2 Age Based Maintenance Policy

In the age based maintenance policy, we assume that the tool is preventively maintained when it reaches a certain number of periods of time N without a failure. Otherwise, it is correctively maintained at the failure time (see Equation 10.14).

$$RM_{t,i}=\begin{cases} 1, & if \quad T_t \geq N \\ 0, & Otherwise \end{cases} \tag{10.14}$$

10.6.3 Age and Availability Based Maintenance Policy

In this policy, PM activity is released when both tools are available and $N \geq T_t \geq n$. In case maintenance is overdue, the activity is released. The formulation of this policy is in Equation 10.15:

$$RM_{t,i} = \begin{cases} PTB_{t,I} < N\text{-}n & \begin{cases} 1\,, & (SM_{t,i}=1 \text{ or } MB_{t,i}=1) \text{ and } AA_t=1 \\ 0\,, & Otherwise \end{cases} \\ PTB_{t,I} \geq N\text{-}n & ,\quad 1 \end{cases} \tag{10.15}$$

Note that when trying to release a backlogged activity, the first thing we do is check whether the time at which the PM scheduled action has been backlogged exceeds the time limits. In the case where we do not exceed the time limits, the maintenance is released only if both tools are OK (*i.e.* $AA_t=1$). In order to facilitate understanding of this PM policy, Figure 10.2 depicts two simple cases of the scheduling and release of PM activities in tool #1, located within a tool set with two tools (T1 and T2), when tool #2 is being maintained. The circles in Figure 10.2 denote the times which trigger backlog, scheduling and release of the PM activity.

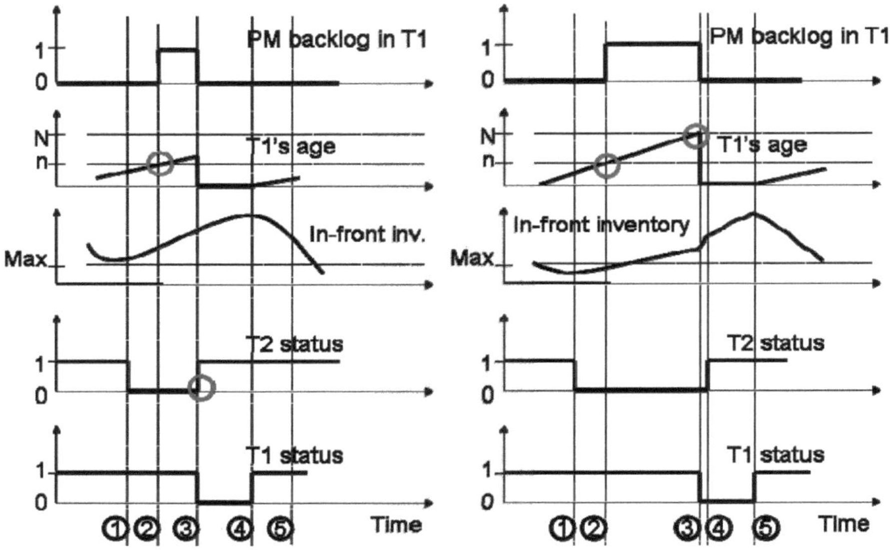

(1) T2 fails;
(2) PM action is backlogged because T1's age reaches n periods but T2 is down;
(3) PM action is released as soon as T2 is up after maintenance and n≤T1's age ≤N;
(4) T1 is up again after the PM, both tools are working and in-front inventory decreases. Note how from (1) to (4) in-front inventory will tend to increase since only one tool is working.

(a)

(1) T2 fails, inventory will tend to increase;
(2) PM action is backlogged, because T1's age reaches n periods and T2 is down. T1's age reached N periods and Pm action is released even with T2 down;
(3) T1 is up again after maintenance, T2 is still down;
(4) T2 is up again. Note how from (3) to (4) the inventory would tend to increase at a higher rate than from (1) to (3) once both tools are down;
(5) T1 is up again after its PM.
Inventory would decrease after this event once both tools would be working.

(b)

Note: In these figures several variables are plotted. With this in mind we would like to make the following remarks. For PM backlog, T2 status and T1 status. Y axis can be either 1 or 0 once they are binary variables describing whether a PM is backlogged or not, T2 is available or not and T1 is available or not, respectively. For in-front inventory Y axis means units of inventory. For T1's age, Y axis means time units.

Figure 10.2. Cases for age and availability based PM Policy. **a** Case 1, sequence of events; and **b** Case 2, sequence of events

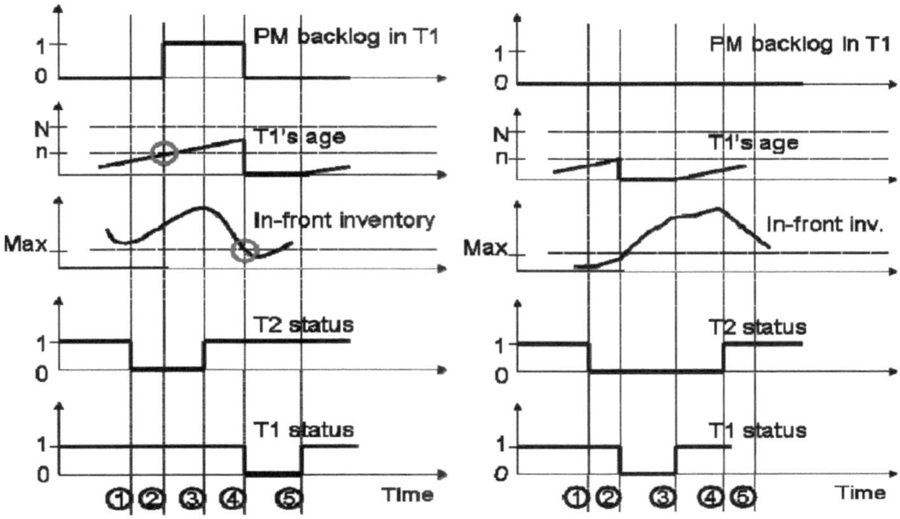

(1) T2 fails;
(2) PM action is backlogged because T1's age reaches n periods but in-front inventory is over maximum value;
(3) T2 is up after maintenance, n≤T1's age ≤N, but in-front inventory is still over the maximum, therefore PM maintenance action keeps backlogged. Note how from (1) to (3) in-front inventory will tend to increase since there is always just one tool working;
(4) T2 is up, n≤T1's age ≤N and in-front inventory is now under the maximum, therefore PM maintenance action is released in T1. Note how from (3) to (4) the inventory will tend to decrease since both tools are working. After (4) inventory would again increase once only T2 would be working.

(a)

(1) T2 fails, inventory will tend to increase;
(2) PM action is released, not backlogged, because T1's age reaches n periods and in-front inventory is below the maximum value. Both tools are down and inventory will increase at a higher rate;
(3) T1 is up again after maintenance, T2 is still down;
(4) T2 is up and both tools are again working. Note how from (3) to (4) the inventory would tend to increase at a similar rate than from (1) to (2). After (4) inventory would decrease once both tools would be working.

(b)

Note: In these Figures several variables are plotted. With this in mind we would like to make the following remarks: For PM backlog, T2 status and T1 status. Y axis can be either 1 or 0 once they are binary variables describing whether a PM is backlogged or not, T2 is available or not and T1 is available or not, respectively. For In-front inventory Y axis means units of inventory. For T1's age, Y axis means time units.

Figure 10.3. Cases for age and in-front buffer based PM Policy. **a** Case 1, sequence of events; and **b** Case 2, sequence of events

10.6.4 Age and In-front Buffer Maintenance Policy

Figure 10.3 shows some examples and explanation of cases when PM activities are released based on tool's age and in-front inventory status (for two tools in parallel in the same tool set processing the same operation).

As we mentioned earlier, maintenance policies explored in this work are based not only on current manufacturing tools status but also on several operating conditions of the wafers production flow. One of these conditions is the WIP status. It seems reasonable that low in-front stock status could be a desirable condition to release maintenance activities since queuing phenomena could be reduce during tools preventive downtime:

$$RM_{t,i} = \begin{cases} PTB_{t,I} < N\text{-}n & \begin{cases} 1, & (SM_{t,i} = 1 \text{ or } MB_{t,i} = 1) \text{ and } IFS_t \leq MxW_i \\ 0, & \text{Otherwise} \end{cases} \\ PTB_{t,I} \geq N\text{-}n, & 1 \end{cases} \tag{10.16}$$

Equations 10.15 and 10.16 are to determine when the *RM* (release maintenance) variable has a value of 0 or 1. For instance, in Equation 10.16, note that maintenance is released ($RM=1$) when:

1. The tool's age is within the maintenance time window ($n<PTB_t<N$), there is a maintenance already backlogged or just scheduled for this time period, and all tools are available;
2. We cannot backlog a maintenance action any longer (because we run out the feasible time window to backlog maintenance (N-n)).

Note that these equations reproduce possible preventive maintenance scheduling policies on the shop floor.

10.7 Specific Wafer Production Flow Scenarios

In order to apply the above maintenance scheduling concepts, the following two simple production flow scenarios are used in our simulation experiments. The production flow Scenario 1 (in Figure 10.4) consist of two tools sets ($TSet_1$ and $TSet_2$) each with only one tool ($Tool_1Tset_1$ and $Tool_2Tset_2$). The production flow scenario 2 (in Figure 10.5) consists of two tools sets ($TSet_1$ and $TSet_2$), the first containing two tools ($Tool_1Tset_1$ and $Tool_2Tset_1$) and the second with only one tool ($Tool_3Tset_2$). TSet1 performs operation 1 and $TSet_2$ operation 2.

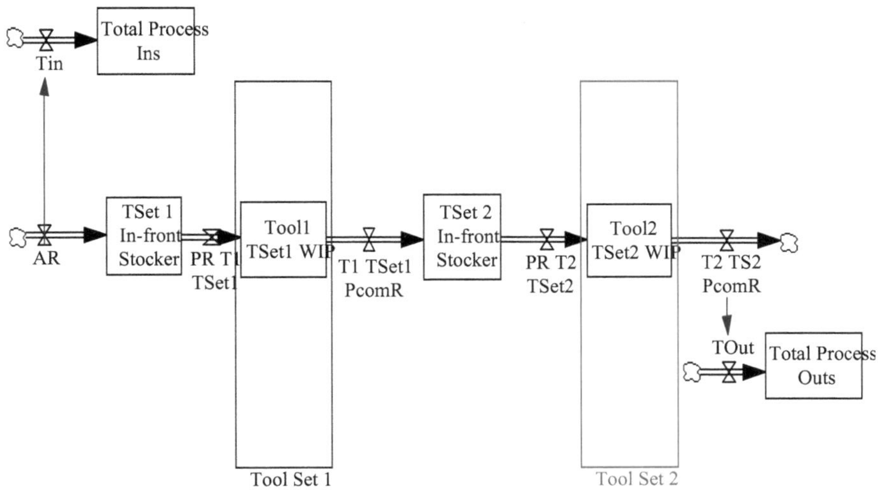

Figure 10.4. Production flow configuration for Scenario 1

In Figures 10.4 and 10.5, the variable *PR Ti* TS_j denotes the production rate of *Tool i* in $TSet_j$. The variable T_i TS_j *TcomR* denotes the production completion rate of *Tool i* in $TSet_j$.

In-front stockers ($TSet_i$ *In-front Stocker*) this time will be capacity constrained (to 200 wafers, which is a low number in real fabs but adjusts reasonable top accumulations for the wafer flows we are considering), and reaching this stockers constraint will provisionally stop the inflow of wafers to those stockers. Tools production capacity will also be constrained to the tools processing speed, which will be according to Table 10.1.

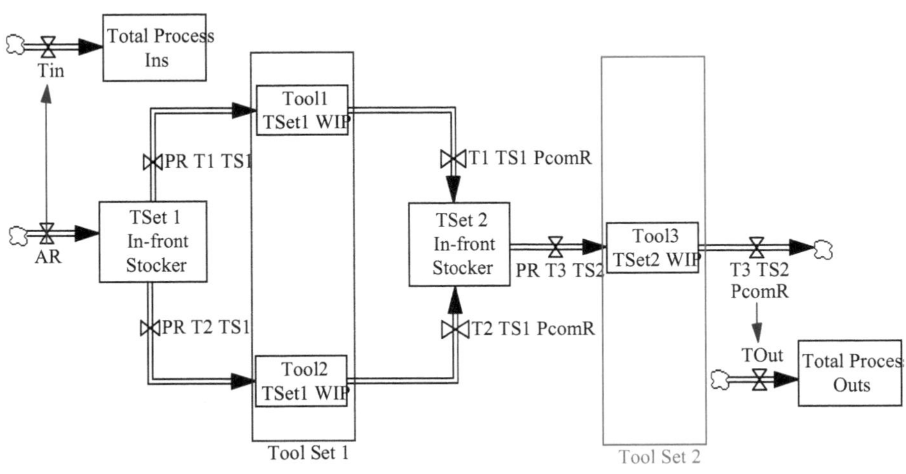

Figure 10.5. Production flow configuration for Scenario 2

In all scenarios, arrival rate of wafers (AR_t in Figures 10.6 and 10.7) is assumed to be exponential (λ=0.428 wafers/min) and failure rate of all tools in all cases is assumed to be uniformly distributed (μ=0.000145 failures/min) within the interval of "time since last maintenance" (age T_t) for which the tool is in operation. Even when preventive maintenance is momentarily delayed, we assume that the same failure rate is applicable. (Note: selection of parameters values was carried out according to real maintenance and repair database records obtained from semiconductor fabs. Initial conditions and simulation horizon were selected in order to test the different policies ensuring that enough number of maintenance actions would be scheduled and released for each simulation.)

Table 10.1. Tool processing time for the scenarios (in minutes)

	Scenario 1	**Scenario 2**
$Tool_1Tset_1$	2	4
$Tool_2Tset_1$	-	4
$Tool_2Tset_2$	2	-
$Tool_3Tset_2$	-	2

For each machine, we will assume that preventive maintenance actions are planned to take place every 1,440 min. Age intervals to explore alternative maintenance scheduling policies will go from n=1,440 to N=(n+1,400)=2,880 min (i.e. there is a range of 1,400 min for a possible PM delay in order to meet more favourable system operational conditions to carry out the PM). With this condition, no more than one maintenance action can be backlogged at any time.

Time required to accomplish a PM is assumed to be 200 min while the unscheduled maintenance (UM) will require 800 min. This will be for all tools and scenarios to simulate. Some readers may argue that this time should also be random; we, however, assume that maintainability programs have taken them to variability ranges where this assumption is reasonable.

In all cases, initial conditions of the system are assumed to be as follows: initial number of wafers in *TSet 1* in-front stocker= 5 wf., initial number of wafers in *TSet 2* in-front stocker =5 wf., initial operational condition for all tools= "idling", initial "time since last maintenance" (age T_0) of the tools 0, 50, and 100 min for tool 1, tool 2, and tool 3 respectively.

In the simulations, as argued in the introduction of the case study, the overall objective against which maintenance scheduling policies are compared will be to maximise the throughput of the wafer manufacturing process (total output) and not the availability of individual tools.

Finally, the simulation horizon will be a total of 40,000 min in all cases. Age based and age and in-front buffer maintenance scheduling policies are tested in Scenario 1, while in Scenario 2 we also test the age and availability policy for tools in $TSet_1$.

10.8 Simulation Results

10.8.1 Introduction to Results of the Case Study

In this section we will discuss the results obtained from our simulation experiments for each scenario. Managerial implications of these results will also be summarised.

10.8.2 Results for Scenario 1

Sample results considering the same set of seeds for the pseudo-random number generation in the model are presented in Figure 10.6. These results deal with the arrival of wafers and the failure appearance stochastic mechanism.

A total of 400 simulation runs were made for maintenance policy classes, grid that could be characterised as *Class Max1, Max2*. Such a maintenance scheduling class would release maintenance of *Tool i* when, within their planned age window [1,440, 2,840], the *in-front stocker i* is below *Max i* value. For this example, there are clear areas of higher performance of the wafer production process (more than 3% output improvement for the optimal policy). At the same time, and for a certain area, simulation results for *Class Max1,Max2* policies are better than those obtained for age based maintenance policy (this is 14,885 wafers), assuming age for PM to be 1,440.

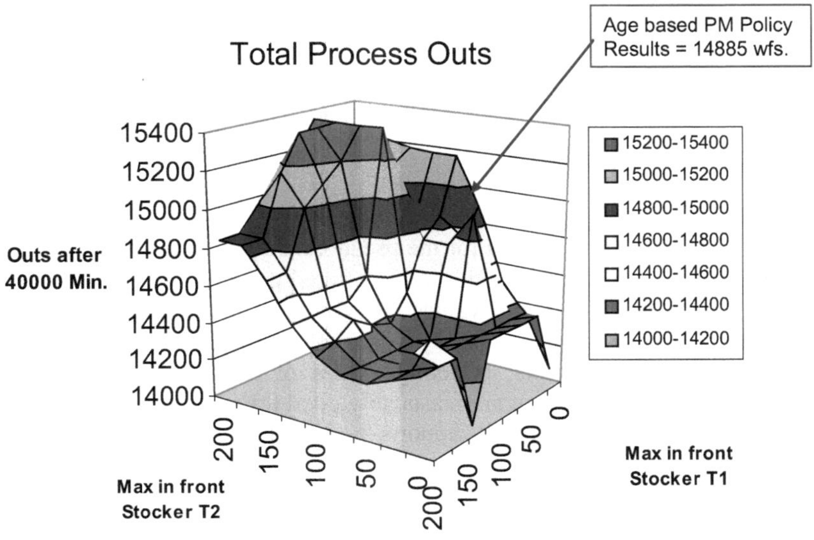

Figure 10.6. Comparison of age based *vs* age and in-front buffer based scheduling policy results for different levels of maximum inventory (Max in Figure 10.3) in both stockers, Scenario 1

Initially, results are a little counter-intuitive. For the cases where the tool failure rate does not increase, we would expect to find policies delaying maintenance as much as possible (*Class 0,0*) to produce better performance (less down time due to lower number of PM along the simulation horizon). However, this is not happening. *Class 77,171* was found to be the best of the high performance classes. Although this is a sample result for a certain set of seeds values, we could simulate other cases (seed sets) and obtain similar results.

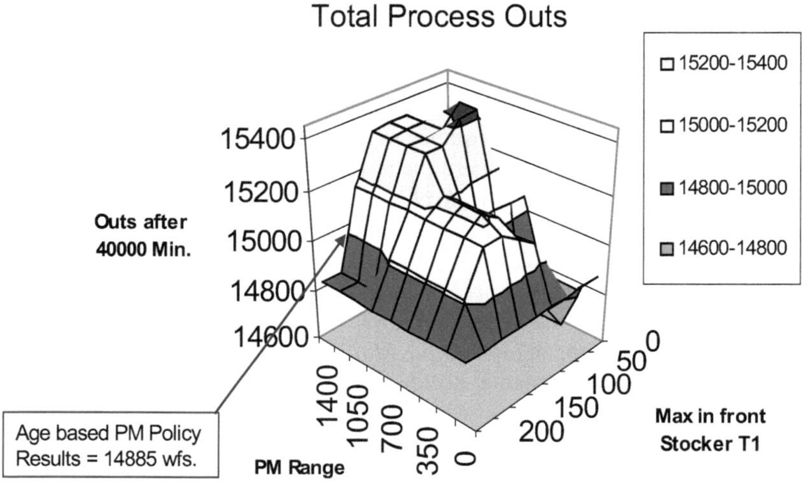

Figure 10.7. Comparison of age based *vs* age and in-front buffer based scheduling policy results for different levels of maximum inventory of stocker 1 and range for the PM (N-n) in Equation 10.17, Scenario 1

Figure 10.7 presents the results corresponding to another 400 simulations considering the same set of seeds for the pseudorandom number as in Figure 10.6. It is fixed *Max 2* =170 wf, since this value offered good performance in previous graph for a wide range of *Max 1* values. The intention was to explore whether by shortening PM range values, the system performance could be maintained. Surprisingly, in this example, selecting values of *Max 1* within the interval [75,150] results for total output are reasonably good for values of PM range (*N–n*), even closer to 300 min. This result supports the idea that too much relaxation of the PM interval does not yield good payback.

10.8.3 Results for Scenario 2

For Scenario 2, Figure 10.8 shows results for *Class Max1, Max2* maintenance policies. We have maintained the same seed of random numbers generation for simulations as in previous figures. Now, policy *Class 124,138* produced the best results, although age and in-front buffer policy offer some lower relative improvements than in previous scenario (around 2% potential improvement for this example). For this scenario, comments regarding the optimal policy resulting in this example are similar to those for Scenario 1 in Figure 10.6.

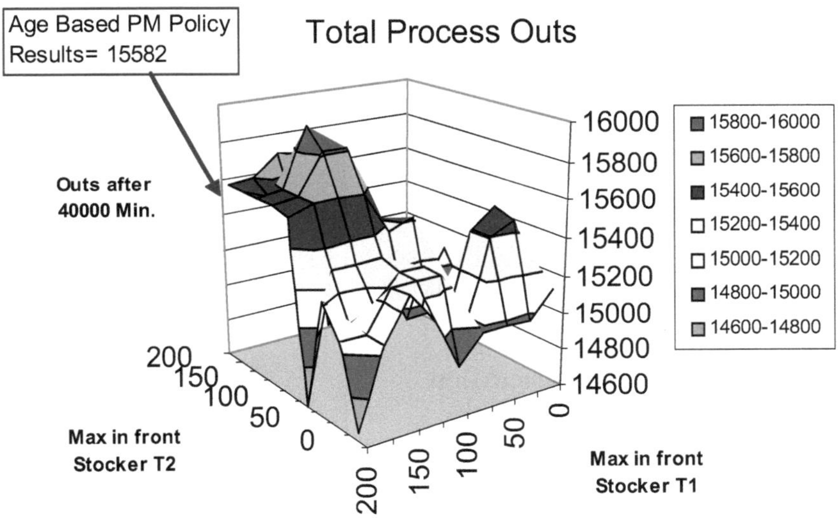

Figure 10.8. Comparison of age based *vs* age and in-front buffer based scheduling policy results for different levels of maximum inventory (Max in Figure 10.3) in both stockers, Scenario 2

For this second scenario, we will now consider the age and availability based scheduling maintenance policy. As mentioned in Section 10.6.3, this policy depends on the range *(N–n)* that allows for maintenance to be delayed in case one tool is down when releasing the scheduled maintenance of a parallel tool. We refer, of course, to tools within the same tool set. In these simulations, we explore the importance of the PM delay range for improving total process output when using this class of policy (*Class N–n, with n=1,440 fixed*).

In Figure 10.9, we present results for three seed sets plus the average of the three. Note that for *N–n*=0, we would be talking about the simple age based policy with *n=1,440*. These results show that increasing the range of maintenance delay improves performance (output) until range is around 1,000 min. For this example, the average performance would be around 1.3% for the optimal policy (*Class 500*).

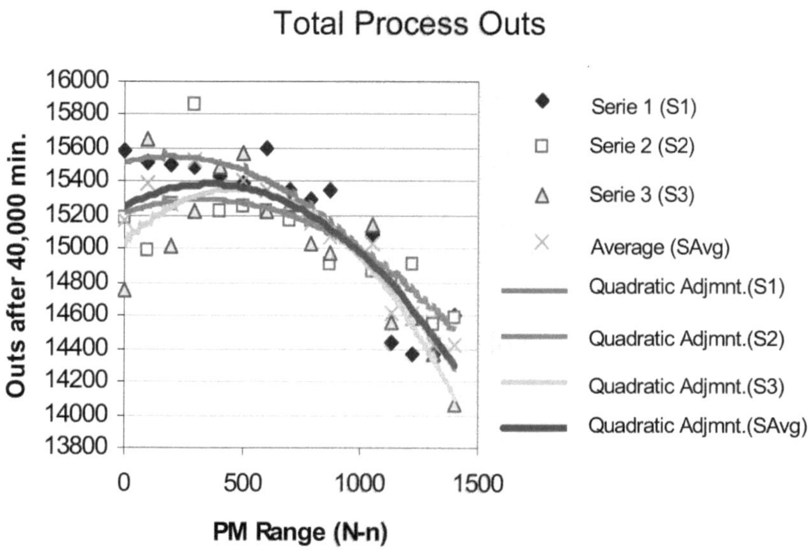

Figure 10.9. Comparison of age and tool availability scheduling policy results for different values of the range for the PM (N–n) in Equation 10.16, Scenario 2

10.8.4 Confidence in Simulation Results

So far we have presented results for a given set of random numbers for the stochastic mechanism of arrival times and failure rates. We will now attempt to build confidence in the results obtained by using some statistical inference techniques applied to results in areas showing important potential improvements. To obtain acceptable insights from these tests, we will constrain this exercise to the number of simulations required. Note that the stochastic mechanism of the mentioned variables is functioning every single minute of the 40,000 min of the planning horizon. This means that we *initially* do not expect big changes by selecting different seeds and adding replications. Of course, results depend on many other factors and we want to be sure about subsequent conclusions we may arrive in this work (see comments about this aspect in [25]).

Results in Table 10.2 for the policies selected are not what we thought they would be. The age and in-front Stocker based policy (AIFSBP) does not perform as expected in both scenarios (1 and 2). We can confirm this impression by looking at Table 10.3, where we can appreciate that age based policy (ABP) performs slightly better that AIFSBP in terms of the mean value for output and expected variability of the output.

Table 10.2. Results obtained for the selected scenario class policies and for 10 replications of the experiments with different seeds (in minutes)

	Process Outs		Process Outs		Process Outs	
		AIFSBP		AIFSBP		AABP
seed	ABP	*Class 77.170*	ABP	*Class 124.138*	ABP	*Class N-n=500*
1	14,802	14,796	15,169	14,435	15,169	15,356
2	15,189	15,182	15,018	14,737	14,919	15,019
3	14,885	15,365	15,582	15,718	15,582	15,466
4	15,170	15,564	15,043	14,564	14,040	15,555
5	14,531	14,125	14,898	15,379	14,749	15,565
6	14,567	14,452	15,370	14,782	15,372	15,423
7	14,988	14,867	14,810	14,852	14,810	15,090
8	14,687	14,471	14,968	14,614	15,180	15,314
9	15,471	14,264	15,239	15,115	14,760	14,907
10	14,657	14,865	15,239	15,115	15,241	15,186
	Scenario 1		Scenario 2		Scenario 2	

Table 10.3. Confidence intervals for policy results in Table 10.2 (in minutes)

					95% Confidence Interval			
Scenario 1	Age Based Policy		Age and In front Stocker Based Policy		Age Based Policy		Age and In front Stocker Based Policy	
Class 77.170	μ	s	μ	s	Min	Max	Min	Max
	14,894.7	307.5	1,4795.1	474.9	14,704.1	15,085.3	14,500.8	15,089.4

					95% Confidence Interval			
Scenario 2	Age Based Policy		Age and In front Stocker Based Policy		Age Based Policy		Age and In front Stocker Based Policy	
Class 124.138	μ	s	μ	s	Min	Max	Min	Max
	15,133.6	233.2	14,931.1	399.3	14,989.0	15,278.2	14,683.6	15,178.6

					95% Confidence Interval			
Scenario 2	Age Based Policy		Age and Availability Based Policy		Age Based Policy		Age and Availability Based Policy	
Class N–n=500	μ	s	μ	s	Min	Max	Min	Max
	14,982.2	431.8	15,288.1	228.6	14,714.6	15,249.8	15,146.4	15,429.8

Results for the age and availability based policy are positive though. We can appreciate an average 2% increase in throughput plus closer bounds for the 95% confidence interval, which may also imply indirectly lower variability wafer output when applying the selected class of this policy. This seems to be a promising result, and prompts us to explore further this type of maintenance scheduling policies.

10.9 Concluding Remarks of the Case Study

In this case study we explored the opportunity to use Monte Carlo continuous time simulation modelling to improve preventive maintenance scheduling in semiconductor fabs. Using this technique, we have showed that we can produce a reasonable assessment of alternative scheduling policies that could be implemented dynamically on the shop floor. Policies considered were based on the status of current manufacturing tools, tools in-front stocker status and tools availability status. We compared and discussed the relative benefits of different scheduling policies in terms of the number of process outputs produced. In order to do, so we treated simulation results as a series of realistic experiments and used statistical inference to reach reasonable confidence intervals of performance parameters.

Based on the simulation results obtained in this chapter, we conclude that the use of age and availability based maintenance scheduling policy (AABP) maximises process throughput and provides better results than those obtained by simple age based maintenance scheduling policy (ABP) for Scenario 2. Results for both scenarios in the chapter also show that scheduling policies based on age and in-front stocker level (AIFSBP) may result in higher variability of the flow while producing no real performance improvement in these particular environments. Nevertheless, setting up a maintenance scheduling policy based on age and in-front buffer levels may not be easy and requires extensive computational efforts that may render it less practical for day to day operations. Future research needs to investigate the implications of the use of AABP for more specific tool set configurations in specific scenarios. Also future research to ascertain the process and steps to follow in articulating a final policy to be given to managers on the shop floor would be beneficial and interesting.

10.10 References

[1] Pidd M, (2003) Tools for thinking. Modelling in management science. 2nd. Edition. Chichester. England: Wiley.

[2] Ignizio JP, (2004) Optimal maintenance headcount allocation: An application of Chebyshev goal programming. International Journal of Production Research, 42(1): 201–210.

[3] Crespo Márquez A, Sánchez Herguedas A, (2002) Models for maintenance optimisation: A study for repairable systems and finite time periods. Reliability Engineering and System Safety, 75(3). 367–377.

[4] Gerstbakh IB, (1976) Sufficient optimality conditions for control-limit policy in a semi-Markov model. Journal Of Applied Probability, 13: 400–406.

[5] Gerstbakh IB, (1977) Models of preventive maintenance. New York: North-Holland.

[6] Gerstbakh IB, (2000) Reliability theory. With applications to preventive maintenance. Berlin: Springer-Verlag.

[7] Papazoglou IA, (2000) Semi-Markovian reliability models for systems with testable components and general test/outage times. Reliability Engineering and System Safety, 47: 175–185.

[8] Becker G, Camarinopoulos L and Zioutas G, (1999) A Semi-Markovian model allowing for inhomogenities with respect to process time. Reliability Engineering and System Safety, 70(1): 41–48.
[9] Abboud NE, (2001) A Discrete-time Markov production-inventory model with machine breakdowns. Computers and Industrial Engineering, 39: 95–107.
[10] Campbell JD, Jardine AKS, (2001) Maintenance excellence: Optimizing equipment life-cycle decisions. New York: Marcel Dekker.
[11] Dekker R and Groenendijk W, (1995) Availability Assessment Methods and their application in Practice. Microelectron Reliability, 35(9–10): 1257–1274.
[12] Scarf PA, (1997) On the application of mathematical models in maintenance. European Journal of Operational Research, 99: 493–506.
[13] Bellman R, (1957) Dynamic Programming. Princeton. New Jersey: Princeton University Press.
[14] Howard RA, (1960) Markov processes and dynamic programming. New York: Technology Press and Wiley Press.
[15] Yao X, Fernandez-Gaucherand E, Fu MC and Marcus SI, (2004) Optimal preventive maintenance scheduling in semiconductor manufacturing. IEEE Transactions on Semiconductor Manufacturing, 17 (3): 345–356.
[16] Hoyland A, Rausand M, (2004) System reliability theory. Models, statistical methods and applications. Wiley Series in probability and Statistics. New Jersey. Hoboken: John Wiley & Sons Inc.
[17] Marseguerra M, Zio E, (2000) Optimizing maintenance and repair policies via combination of genetic algorithms and Monte Carlo Simulation. Reliability Engineering and System Safety, 68: 69–83.
[18] Hopp WJ, Spearman ML, (1996) Factory physics. Foundations of manufacturing management. Chicago: IRWIN.
[19] Hopp WJ, Pati N, Jones PC, (1989) Optimal inventory control in a production flow system with failures. International Journal of Production Research, 27: 1367–1384.
[20] Hsu LF, (1999) Simultaneous determination of preventive maintenance and replacement policies in a queue-like production system with minimal repair. Reliability Engineering and System Safety, 63(2): 161–167.
[21] Liu B, Cao J, (1999) Analysis of a production-inventory system with machine breakdowns and shutdowns. Computers and Operations Research, 26: 73–91.
[22] Simon JT, Hopp WJ, (1995) Throughput and average inventory in discrete balanced assembly systems. IIE Transactions, 267: 368–373.
[23] Van Der Duyn Schouten FA, Vanneste SG, (1995) Maintenance optimisation of a production system with buffer capacity. European Journal of Operations Research, 82: 323–338.
[24] Crespo Marquez A, Gupta JND, Sánchez Herguedas A (2003) Maintenance policies for a production system with constrained production rate and buffer capacity. International Journal of Production Research, 41 (9): 1909–1926.
[25] Andijani A, Duffuaa S, (2002). Critical evaluation of simulation studies in maintenance systems. Production Planning and Control, 13(4): 336–341.
[26] Marseguerra M, Zio E (2002) Basics of the Monte Carlo method with applications to system reliability. Hagen. Germany: LiLoLe-Verlag GmbH.
[27] Vensim, Version 5.4. (2004) Ventana Systems inc. Harvard. Massachusetts.

Part IV

Modelling Integration Issues in SCM

11

Different Supply Chain Integration Models

11.1 SC Integration Opportunities

11.1.1 Overview

As mentioned in Chapter 2, when considering planning and control activities, the effectiveness of SC integration is important and this may depend on the tools used for SC integration, and on the sequence of implementation of these tools.

Typically, the four main integration phases are:

1. information sharing, including r&d information for product design and the information to track the materials flow along the chain;
2. collaboration for a common forecast;
3. common planning; and
4. automated financial transactions.

These phases could be implemented with a different sequence. For instance, the information about the material flow could be used to plan the build rates along the chain. However, in such a case, a common forecast would not be accessible. At the same time, the partners of the chain could have access to a common forecast, but the available inventory information would only be local. Thus, global inventory information would not be used. A visit to existing internet portals for e-collaboration indicates that automated financial transactions could be introduced at different points of the process.

Based on their use, the e-collaboration supply chain integration tools can be categorized into the following five classes [1]:

1. tools to "wire" the company, offering real time information about the material flow, which is basically managed by exception;
2. tools to share documents in real time;
3. tools to do collaborative forecasting;
4. tools to do collaborative planning (currently very scarce);
5. tools to implement automated payments (currently very scarce).

But in order to contextualize the problem of SC integration, and prior to the presentation of the modelling aspects in our next section, let us review a case study dealing with e-collaboration tools, their scope and potential improvements through the use of dynamic modelling tools.

11.1.2 The Factory.com Case Study

Factory.com was founded and led by electronics industry veterans. Previously, these executives led strategic outsourcing initiatives, but then suffered the operations management challenges of limited information flow throughout their production supply chain.

By the year 2000, Factory.com marketed the "Factory Network" (FN), a product that was introduced to potential customers as the world's first Collaborative Manufacturing Execution (CME) platform, a breakthrough solution for managing *today's* extensively outsourced production operations. Accessed via the Factory.com web site, the Factory.com CME platform could provide secure and seamless functionality to manage production across a collaborative manufacturing supply chain:

- material tracking and management;
- forecast and loading visibility;
- manufacturing data collection; and
- analysis and reporting tools.

Whether integrated with a given in house ERP or MES, or functioning as sole production management system, the Global Factory Network could provide complete visibility to outsourced production processes, and/or to the processes provided to customers.

The Factory Network was designed to integrate easily with in-house and legacy systems, recognising their role as the origin or destination of much operations management data and transactions. Smaller manufacturers could choose to use the Factory Network as their in house operations management solution, in addition to its role in linking them to customers and suppliers.

The revolutionary strength of a shared hub-based CME platform was presented as *the network effect*:

> *"...you only have to model your operation once, and can then easily and near instantly link your production tracking with other Factory Network members and even with non-members. There is no need to set up difficult and expensive point-to-point linkages with each customer and supplier. Simply point and click within the Factory Network, and you can do business your way, while enjoying the benefits of an integrated system across your entire collaborative manufacturing supply chain."*

Setting up a company as a member of the Factory Network was as simple as drawing a flow chart and filling in a few blanks. The Factory Network's field configurable controls were radically easy, allowing a tailoring of the system to specific terminology, data fields, business rules and reporting requirements. The fact that this level of tailoring could be performed by non-information technology personnel meant that not only your company could quickly come up on the Factory Network, but so could your customers and suppliers.

11.1.3 How the Factory.com CME Works

The CME platform operated via shared models of collaborative manufacturing operations. Members could use a flow chart editor to create the models, after which they could provide selective access to their customers and suppliers. This selective access allowed workspace owners to segment information based on a right to know basis. Members could then include other members' model elements into their own collaborative manufacturing models. The result is an integrated multi-enterprise management system.

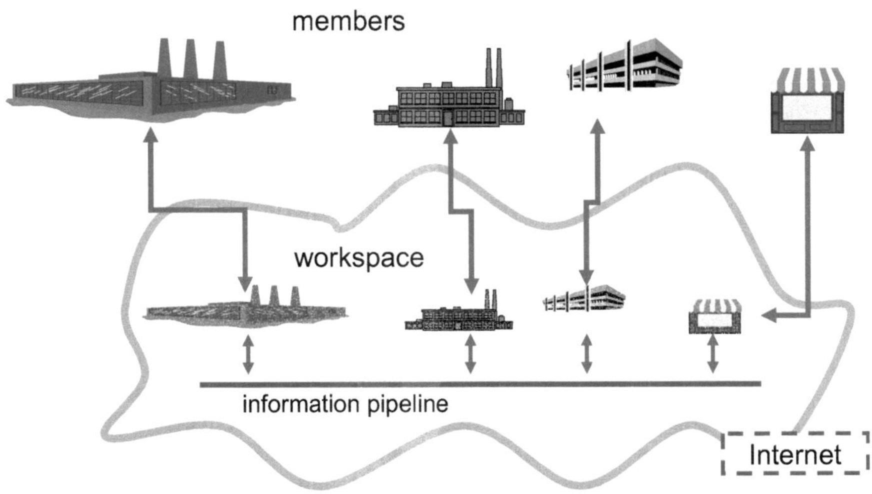

Figure 11.1. Production flow configuration for scenario 1

As more and more companies joined, the Factory Network became a universe of connected enterprises. Each member had a "workspace", a secure area for models and data. Models were composed of "stations" and "flows". Stations represented locations where inventory resides, and typically where fabrication and/or testing operations were performed. Flows were the routes along which material could move between stations. Model properties were user defined, and represented the information to be collected in a station or processing step.

Once granted access to the relevant stations within other member's workspaces, assembling a multi-enterprise management system was as simple as drawing lines between the stations, and identifying the data items to transfer from station to station. The system could be put into immediate usage. Material tracking and manufacturing data collection activities that were recorded in the Factory Network were automatically propagated down the supply chain.

FN offered four biz process modules:

- *Multi-Enterprise WIP and Order Flow Manager*: this module ensured that materials and orders were not only processed correctly and on schedule, but that the controls and quality were in place to ensure customer satisfaction.

  The Work In Process (WIP) and Order Flow Manager allowed real time material tracking in a multi-company production environment. It provided the necessary visibility to ensure that all material was represented correctly and moving according to the production plan. It allowed complex material tracking operations, such as splits or combines, with full traceability of the origin.

  A full set of business rules could be applied to allow continuous and accurate representation of material location. For example, even if company "A" represents material as "shipped" while company "B" has already received it, the system would show the material in the right location so no double counting occurs. In every step the system collected all the relevant engineering and QA parameters to ensure product quality and to support process and yield improvements. Together with other powerful tools in the Factory Network, such as alerts, exception reporting and data analyses, this module ensured that materials and orders were not only processed correctly and on schedule, but that the controls and quality were in place to ensure customer satisfaction.
- *Multi-Enterprise Document Sharing*: this allowed all partners to operate from the same drawings, process, and product documentation. It provided secure and seamless real time document routing capabilities, so that product and process designers could easily and effectively deliver current and accurate instructional and reference material to partners. Instead of sending drawing, specs, engineering data and other documents back and forth, not knowing if everyone was using the most current revisions, this module allowed everyone to look at the latest revisions. It also managed different view and edit permission levels for different users at different companies.
- *Collaborative Forecasting*: the Collaborative Forecasting module was used by customers and suppliers jointly to communicate and refine production forecasts. The system allowed each member to maintain their forecast in their own terminology, and with their own part numbering scheme. Built-in graphical tools tracked forecast accuracy and provided aggregated forecasts for optimum collaborative synergy.
- *Collaborative New Product Introduction Manager:* used to manage and communicate multi-company data flows associated with the details of new products. For instance, it allowed OEMs, assembly houses and materials suppliers to collaborate on the availability of a new material, the inclusion of that new material into a new assembly, and the rate at which the new assembly could be brought up to full production levels. The result was faster, more error-free new product introductions.

11.1.4 The FN Architecture

The Factory Network was comprised, as mentioned above, of four business process modules, which were surrounded by a comprehensive set of administration modules and pervasive business intelligence features. The entire system was based on a Secure Collaboration Provider infrastructure as shown in Figure 11.2.

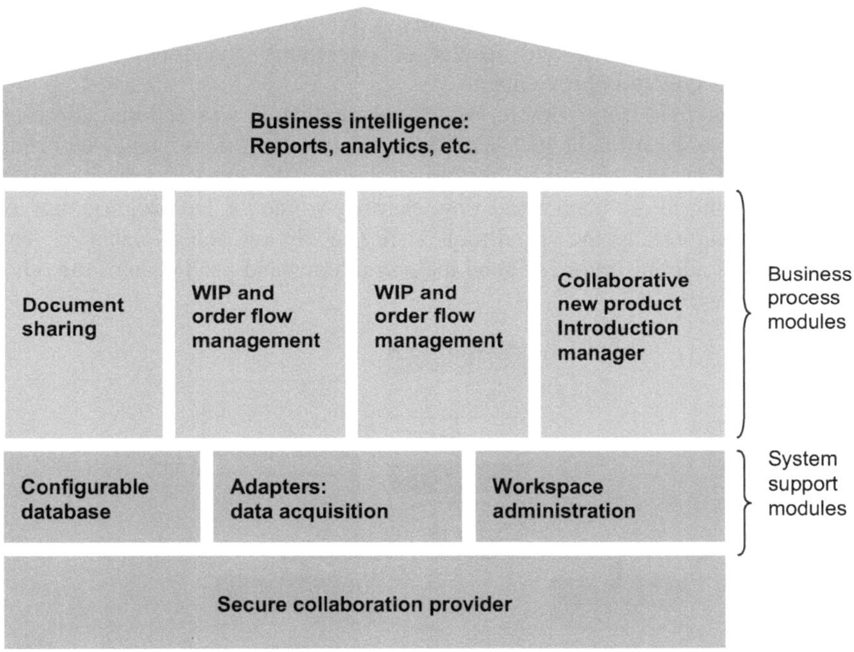

Figure 11.2. Details of the CME platform

The system's *Secure Collaboration Provider* infrastructure delivered the mission critical reliability, security and scalability required for collaborative manufacturing execution management. This architecture allowed for automatic propagation of information, decisions and results across the universe of Factory.com workspaces while providing each member absolute control over their proprietary information. Additionally, Factory.com deployed a variety of optimisation strategies to provide maximum responsiveness and efficient use of resources including:

- replication of data across workspace boundaries for optimised access where it was needed based on right to know;
- unlimited processing and network resources to be allocated to any member – all bottlenecks could be readily avoided;
- optimisation of both the transactional system and rapid data retrieval;
- support of fully secure sessions while maintaining responsiveness;
- workspace level availability of solid state databases: maximising performance.

11.1.5 Business Intelligence, Configuration Tailoring and Integration

The FN included system wide business intelligence features, all of which were end user accessible without requiring IT involvement. These included user definable reports, including exception reporting. The system also included a comprehensive alerting system, allowing stakeholders throughout the supply chain to be immediately alerted to out of range production, quality or logistics results anywhere in the supply chain. Business analytics supported better understanding and analysis of the status and results of operations and provided a clear understanding of how to improve them.

The FN was field configurable, which meant that it was tailored for each member's unique terminology, workflow, data definitions and reporting requirements. This tailoring was performed in the Workspace Administration modules by filling in the blanks and flow charting windows. No programming or scripting was required, as the visual tools were friendly and self explanatory. The user assigned as administrator assigned the permissions and privileges of the other users on the system.

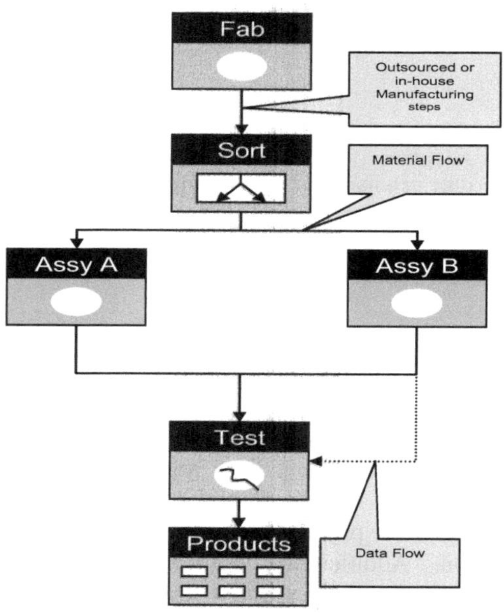

Figure 11.3. Details of the FN CME platform

Each member of the FN configured their system to their own requirements (see Figure 11.3). This capability was based on the configurable database technology which Factory.com developed. This was a "patent pending technology" which at that time allowed rapid implementation of the system, without incurring major IT resource allocation. Moreover, the database could be reconfigured time and again as requirements changed.

With regards to adapters and integration, in the FN resided top internal enterprise applications, such as ERP and MES. The FN Adapter was a comprehensive system integration module used to exchange data easily between FN and these internal enterprise applications. The Adapter provided an open interface, with wide format support, including standards such as RosettaNet. The adapter control was a set of monitoring and management functions for connections between the Factory.com system and other systems. The Adapter could be configured for the usage of available adapters, such data communication channels between an adapter and a target system, data mapping between the systems, etc. There was a library of existing available adapters, including standard adapters for HTML, Excel and XML data transfer, and additional adapters for various ERP and MES systems, such as Oracle Applications, SAP, Consilium, CamStar, etc.

11.1.6 Partnering Options with Factory.com and Modelling Opportunities

By the year 2000, Factory.com was offering the following partnering options to other companies:

- *VAR Partners*: Value Added Reseller (VAR) partners were certified to sell and install factory.com solutions and as such provide both pre- and post-sale support services and could also provide application interface services.
- *Implementation Partners*: Implementation partners were certified to provide consulting services for Factory.com customers. The services could include, installation of Factory.com products, application interface services, and custom integration to existing software. Implementation partners could choose to operate in a pure consultancy partnership with FN.
- *Technology Partners*: Factory.com aligned with the industry's leading technology providers to take advantage of the latest software and hardware products with features that could complement Factory.com's software.

During the year 2000, modelling opportunities appeared at Factory.com. A modelling team was studying the potential benefits of dynamic simulation as a full complementary tool for the FN development. The following models were found to be of major interest for Factory.com:

- *Multi-Enterprise WIP and Order Flow Dynamic Simulation Models*: these models were built in such a way to allow the easy scalability of nodes in the network. Basically they contained all variables related to the material flow: production rates and inventory (on hand, in process or in transportation) accumulation, plus the variables related to the information flows, such as forecast and production plan, capacity constraints, allocation procedures, ... etc. And at the same time, variables related to the financial flows could be attached (working capital requirement of the different nodes in the network, in order to fulfil a certain market demand, could be estimated over time, and study the implications on financial aspects on information flow changes). The variables defined as part of the information flow could determine the production/supply orders in the different nodes of the collaborative network and therefore drive the velocity and flexibility of the system. By changing the structure of the information flows in the model, the impact on the system was easily appreciated, and the benefits of potential actions could be estimated. In fact, design of the information flow according to the network capabilities and the environmental/market conditions could be improved.

- *Collaborative Forecasting Modelling*: this model could be implemented within the previous one, the idea was to examine the implications of a collaborative forecasting in a supply chain. Four possibilities for collaboration levels in terms of forecasting were analysed:
    - *Independent decision-making.* It was assumed there was no communication or collaboration between nodes. The only information that is passed on concerned orders and delivery performance. Each node makes their forecast and places its orders accordingly. There is no trust among nodes and ordering is a way to show discontent or influence supplier behaviour.
    - *Sharing sell-through.* This model considers that each node still makes its own forecast and orders accordingly. Each node trusts its upstream partners to do the right thing with this information, but these partners are still dependent on each other's behaviour to exploit this information.
    - *Sharing sell-through and collaborative planning.* Demand information is in this case passed to the upstream network nodes, and all the nodes use the same forecast to place their orders. The network now collaborates on meeting end-customer demand, and discusses issues and sales expectations on a weekly basis.
    - *Network integration.* In this model, collaborative planning would extend to inventory management and ordering in the entire network.
- *System and model for Product Investments Selection*: this model could be used to determine the most suitable product investment to accomplish in order to penetrate a given market. The model could be configured for multiple markets (segments), products, competitors and product attributes. This model has been used with enormous success within several Hi Tech companies.
- *Model for product introduction analysis*: this model explored the impact of technology credibility and awareness in the process of a product introduction. It helped to understand the timing of this process and could be parametrised for different products/models with different sequences of introduction, regions, divisions, etc. The model could also study the financial implications regarding different metrics, of the market penetration process.

11.2 Characterisation of SC Materials and Information Flows

11.2.1 Material and Information Variables

In order to proceed with the model development and discussion, we will first describe the notations and definition of the main variables as follows (please note that the explanation for each of these variables will be given later in the Chapter):

Information related variables:

D_t^{i+1} Orders of units received in the node *i* in period *t*

DC_t^{i+1} Orders received in the node *i* in period *t*, when node *i*+1 has financial constraints

B_t^i Existing backlog of orders in node *i* in *t*

S_t^i Amount of orders finally shipped to the next node *i*+1 (equivalent to units shipped to the next node) in *t*

$\hat{\mu}_t^i$ Forecast of node *i* in period *t*

ib_t^i Information provided to the node *i* through the information backbone in time *t*

Material related variables:

P_t^i Pipeline from node *i* to the next node *i*+1, (includes work in process inventory in the node plus the inventory of parts in transportation to the warehouse of finished materials) in period *t*

Y_t^i Finished materials inventory of the node *i*, on-hand inventory in period *t*

S_t^i Amount of units finally shipped to the next node *i*+1

O_t^i Output from the pipeline of node *i* in *t*

I_t^i Input to the pipeline of the node *i* in *t*

Model parameters:

L^i Lead time for a material unit in the pipeline to arrive to the inventory of finished materials

ss^i Desired time for a material unit to remain as on-hand inventory of node *i* (this is a policy of each node)

α^{i} Node i forecast smoothing factor

β_{S} Fractional adjustment coefficient for the on-hand inventory

β_{SL} Fractional adjustment coefficient for the pipeline inventory

11.2.2 Characterisation of SC Materials and Information Flows

In our model, it is assumed that the orders received at node i, D_t^{i+1}, are immediately shipped, up to availability, to node $i+1$. When delivering materials, inventory constraints may appear at the node reducing the amount of units finally shipped to the next node, S_t^i (see Figure 11.2). The equations for the orders delivered are as follows:

$$S_t^i = \begin{cases} B_{t-1}^i + D_t^{i+1}, & \text{if } Y_t^i \geq B_{t-1}^i + D_t^{i+1} \\ Y_t^i, & \text{if } Y_t^i < B_{t-1}^i + D_t^{i+1} \end{cases} \tag{11.1}$$

Equations 11.2 and 11.3 show the calculations for the level of backlog and on-hand inventory:

$$B_t^i = B_{t-1}^i + D_t^{i+1} - S_t^i \tag{11.2}$$

$$Y_t^i = Y_{t-1}^i + O_t^i - S_t^i \tag{11.3}$$

Equation 11.4 expresses the input to the pipeline of node i as the shipments from node i–1:

$$O_t^i = I_{t-L^i}^i \tag{11.4}$$

Equation 11.5 formalises the output of the pipeline from node i as a delay of time L^i of its input:

$$I_t^i = S_t^{i-1} \tag{11.5}$$

Equation 11.6 shows the calculation of the pipeline inventory:

$$P_t^i = P_{t-1}^i + I_t^i - O_t^i \tag{11.6}$$

All previous relationships are shown in Figure 11.4, where the dotted lines indicate information flows and the continuous lines show the material flows.

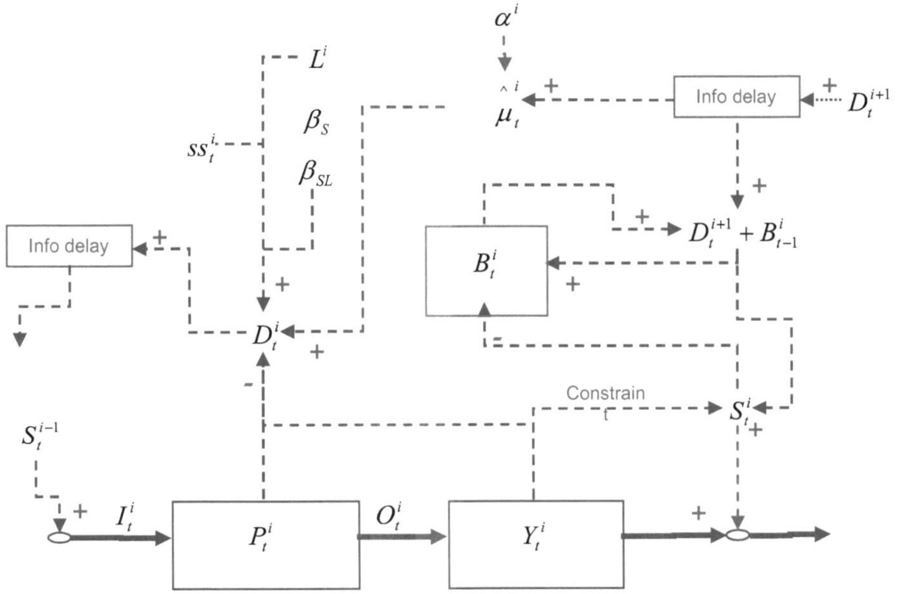

Figure 11.4. Basic influence diagram for the variables in node [i], with no SC integration

11.2.3 Modelling Information Flows According to the Integration Sequence

Table 11.1 depicts various integration possibilities in a supply chain. In the first case, there is a non-integration in the SC (called non-integration – NI – in Table 11.1), meaning that there is no communication at all between the nodes (e.g. retailers do not talk to anyone else; the same for wholesalers, distributors, and factories). This is a very common circumstance in real life, when for example, there may be 2 or 3 factories, 20 or 30 distributors, 2,000 or 3,000 wholesalers, and 20,000 or 30,000 retailers. They never find out what the total activity of the others is. Each node produces its own forecast and places its orders accordingly. Therefore, communication is only through orders.

Now, assume that the supply chain is "wired" (i.e. the different members can receive real time information about the materials flow and orders flow along the SC). Then, an option would be to implement tools to carry out *collaborative forecasting* (called partial integration A – PIA – in Table 11.1). In this case, the final SC member would trust its upstream partners to do the right thing with their end-customer's information, and all nodes in the SC would use the same forecast to place their orders. The chain now collaborates on meeting end-customer demand and discusses issues and sales expectations (on a time period/week basis).

Once the SC is wired, another possibility, a second option for the members would be to use the real time information concerning the materials flow, before discussing and sharing any common forecast, and do their planning and inventory management accordingly (called partial integration B -PIB- in Table 11.1).

Finally, the SC may reach a situation where all partners gain total access to information they do not control, concerning end-customer demand and materials flow, and they use it in their planning process. There is no need for local forecast and collaborative planning extends to inventory management and ordering in the entire network (called Full integration – FI – in Table 11.1).

Table 11.1. Integration possibilities

	(NI) No Integration	(PIA) Partial Integration A	(PIB) Partial Integration B	(FI) Full Integration
Demand Forecast	Local	Shared	Local	Shared
Inventory information and planning	Local	Local	Shared	Shared

In order to obtain the mathematical formulation for placing orders to the upstream node, we consider four different possible levels of implementing e-collaboration tools shown in Table 11.1. Thus, we obtain the relationships that we present in the following sections.

11.3 Modelling a Non-integrated Supply Chain

Now mathematical equations for an NI SC will be formalised. In Equation 11.7, an exponential smoothing constant is used to produce the forecast, since it is widely used in modelling a SC (see e.g. [2]), and has been found to be a very popular practice [3]. To choose appropriate values of α , the reader is referred to Makridakis *et al.* [4]:

$$\hat{\mu}_t^i = \alpha^i D_{t-1}^{i+1} + (1-\alpha^i)\hat{\mu}_{t-1}^i \quad \text{with} \quad 0 < \alpha^i \leq 1, \forall i \tag{11.7}$$

$$D_t^i = Max(\hat{\mu}_t^i + \beta_S(\hat{\mu}_t^i ss_t^i - Y_t^i) + \beta_{SL}(\hat{\mu}_t^i L^i - P_t^i), 0) \tag{11.8}$$

Orders placed are modelled using an anchoring and adjustment heuristic (as in Tversky and Kahneman, [5]), which has been shown to apply to this kind of SC decision-making task [6].

11.4 Modelling PI SC with Sharing Sell-through

The forecast of node i in period t is formalised as follows:

$$\hat{\mu}_t^i = \hat{\mu}_t^n, \forall i = 1,..n \tag{11.9}$$

$$\hat{\mu}_t^n = \alpha^n D_{t-1}^{cust} + (1-\alpha^n)\hat{\mu}_{t-1}^n \text{ with } 0 < \alpha^i \leq 1, \forall i \tag{11.10}$$

Where D_{t-1}^{cust} is the last time period demand for the SC end-customer. Once the new node forecast is obtained, the orders are calculated as in Equation 10.8.

11.5 Modelling PI SC with Shared Inventory Information

For this case, while Equation 11.7 is valid, Equations 11.9 and 11.10 are not applicable. Assuming that we know information about all the nodes, and it is in the backbone, we use a generalised form of the anchoring and adjustment heuristic [6] in an iterative way. Discounting the backlog from the order since the previous node is already expecting this to ship that amount ASAP and noting that the order quantity cannot be negative, Equation 11.11 replaces Equation 11.8:

$$D_t^i = Max\,(\hat{\mu}_t^i + \hat{\mu}_t^i(ss_t^i + L^i) - (P_t^i + Y_t^i) - B_{t-1}^{i-1} + ib_t^i, 0) \tag{11.11}$$

where ib_t^i is a variable expressing the information provided to the node i through the information backbone in time t:

$$ib_t^i = \hat{\mu}_t^i(ss_t^{i+1} + L^{i+1}) - (P_t^{i+1} + Y_t^{i+1}) + ib_t^{i+1} \tag{11.12}$$

As an additional improvement, the backlog of the upstream node at the end of last period t–1 is included in Equation (10.11) for the orders to be placed by node i. The reason for this is to enable immediate shipments (zero expected backlogs under normal conditions). Furthermore, each node includes its last period backlog as part of the desired shipments in the next period t. Therefore the backlog will be fulfilled as soon as on-hand inventory becomes available.

11.6 Modelling Integrated (Sales and Inventory) Supply Chains

Equations 11.9 and 11.10 are applicable (replacing Equation 11.7), and also Equations 11.11 and 11.12 (replacing Equation 11.8).

Table 11.2. Structure of SC and table for parameters

	Supplier →	Factory →	Distributor →	Retailer	
OPERATIONAL					
L^i	2	2	2	2	week
ss^i	3	3	3	3	week
β_S	0,24	0,24	0,24	0,24	1/week
β_{SL}	0,1	0,1	0,1	0,1	1/week
α^i	0,3	0,3	0,3	0,3	dimensionless
FINANCIAL					
Pm_t^i	500	600	700	1000	$/Unit
$Wso(i)$	2	2	2	2	days
Cm_t^i	0,13	0,13	0,13	0,13	%

11.7 Results About Integration Sequence Implications

In order to apply different SC integration possibilities (NI, PIA, PIB, FI) to a well-known example, a four-node SC (Factory, Distributor, Wholesaler and Retailer) as described by Sterman [7] was selected and modelled. Fractional adjustment coefficients are assumed to be the mean values of the experiments presented in [6] (see Table 11.2). The non-integrated SC structure in this Chapter is intended to be a representation of the scenario described by Sterman [6]. The only exception in this chapter is the assumption regarding the customer's behaviour: the retailer does not hold any backlog (end-customers do not wait). The simulation runs are for a total of 52 weeks.

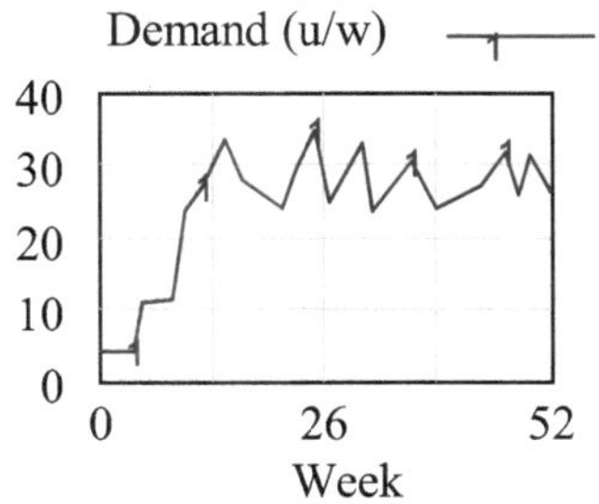

Figure 11.5. End-customer's demand in units per week

In our example, we will change the demand of the retailer; SC end-customers demand (see Figure 11.5, showing a selected real demand of a product). Also we considered a price structure as in Table 11.2, and the price of the raw materials at the factory to be $ 200/unit.

Table 11.3. Scenario 1 results for orders placed (units per week)

Variable **Orders placed (D)**	Min	Max	Mean	StDev
(NI, Factory)	4	74	33	24
(NI, Distributor)	0	60	28	19
(NI, Wholesaler)	0	48	23	16
(NI, Retailer)	1	43	26	15
(PIA, Factory)	4	40	30	12
(PIA, Distributor)	0	40	27	14
(PIA, Wholesaler)	0	40	25	14
(PIA, Retailer)	4	40	25	11
(PIB, Factory)	0	70	27	25
(PIB, Distributor)	0	56	26	16
(PIB, Wholesaler)	4	42	25	11
(PIB, Retailer)	4	37	24	9
(FI, Factory)	4	58	29	15
(FI, Distributor)	4	53	28	13
(FI, Wholesaler)	4	45	26	11
(FI, Retailer)	4	35	24	9

Table 11.3 shows the orders placed by various nodes for different SC integration structures. The results show a higher "bullwhip effect" (amplification in the range of Min-Max orders placed in the first nodes of the SC) for the NI and PIB structures, and a good performance of the PIASC (even better than the FISC). Collaborative forecasting, and therefore the speed of the demand information flow along the chain, turns out to be the most relevant factor conditioning this

amplification problem in this simulation. This is an important result to take into consideration since "bullwhip effect" can be incremented by ordering periodically (batching), by customers overreaction anticipating possible shortages, or by price fluctuations [8].

Table 11.4 shows the SC integration improvements in terms of inventory by node. From these results, it is clear that full integration of the SC results in less standard deviation (StDev) of the units in inventory, although other structures (NI, PIA) yield better mean values.

Table 11.4. Results for inventory of the node (units)

Variable **Total Inventory (P+Y)**	Min	Max	Mean	StDev
(NI,Factory)	11	268	69	52
(NI,Distributor)	11	300	70	68
(NI,Wholesaler)	11	260	68	70
(NI,Retailer)	11	229	66	64
(PIA,Factory)	20	138	61	25
(PIA,Distributor)	20	243	73	54
(PIA,Wholesaler)	17	229	77	58
(PIA,Retailer)	11	174	74	51
(PIB,Factory)	20	309	150	105
(PIB,Distributor)	18	216	96	59
(PIB,Wholesaler)	15	116	68	34
(PIB,Retailer)	11	135	75	44
(FI,Factory)	20	116	87	34
(FI,Factory)	20	127	86	37
(FI,Distributor)	20	127	83	39
(FI,Wholesaler)	11	118	77	40

Table 11.5 shows backlog per node, where the FISC presents the best values, especially for the retailer, in terms of mean, maximum (Max) and standard deviation (StDev).

11.8 Concluding Remarks

Issues involved in the integration of supply chain through the use of e-collaboration tools were considered. A comprehensive model to study the operational benefits of using various e-collaboration tools was developed. A System Dynamics based simulation was used to study the impact of various levels of supply chain integration. Computational results from our experiments clearly show the potential improvements of the integration by using Internet tools for SC collaboration. The sequence of implementing this new technology should start by addressing the issue of collaborative demand forecasting (PIASC), and then continue with the collaborative planning by sharing and using the inventory information of the whole SC (FISC). Implementation of e-collaboration tools to do local planning using global SC inventories data when each node is producing its own forecast (PIBSC) could lead to significant increases in inventory and decreases in income, especially where nodes are not financially constrained.

Full integration of supply chain (FISC) clearly provides more benefits than any partial integration of supply chain. FISC benefits from the visibility of the total materials flow and backlog orders along the chain. As well as the fast access to demand information, it also enables the ordering policies to adjust to new customer requirements earlier and with more efficient inventory administration (less inventory cost to reach a target service level) along the chain.

Table 11.5. Results for backlog of the node (ordered units)

Variable **Backlog (B)**	Min	Max	Mean	StDev
(NI, Factory)	0	105	49	38
(NI, Distributor)	0	152	66	54
(NI, Wholesaler)	0	166	68	61
(NI, Retailer)	0	379	248	153
(PIA, Factory)	0	66	38	26
(PIA, Distributor)	0	95	44	38
(PIA, Wholesaler)	0	86	29	32
(PIA, Retailer)	0	204	149	81
(PIB, Factory)	0	23	3	6
(PIB, Distributor)	0	36	5	10
(PIB, Wholesaler)	0	46	11	16
(PIB, Retailer)	0	190	141	76
(FI, Factory)	0	32	5	10
(FI, Distributor)	0	34	5	11
(FI, Wholesaler)	0	25	3	7
(FI, Retailer)	0	141	107	55

11.9 References

[1] Crespo Marquez A, Bianchi C, Gupta JND, (2004) Operational and financial effectiveness of e-collaboration tools in supply chain integration. European Journal of Operations Research, 159(2): 348–363.

[2] Chen F, Drezner Z, Ryan JK, Simichi-Levy D, (1999) The bullwhip effect: Managerial insights on the impact of forecasting and information on variability in a supply chain. In: Mayur, Ganeshan, Magazine (Eds.), Quantitative Models for Supply Chain Management. International Series in Operations Research and Management Science, 17: 419–439.

[3] Sanders NR, Manrodt KB, (1994) Forecasting practices in US corporations. Survey results. Interfaces, 24(2): 92–100.
[4] Makridakis S, Wheelwright S, Hyndman R, (1998) Forecasting Methods and Applications. New York: John Wiley and Sons.
[5] Tversky A, Kahneman D, (1974) Judgement under uncertainty. Heuristics and biases. Science, 185:1124–1131.
[6] Sterman JD, (1989) Modelling managerial behaviour: Misperceptions of feedback in a dynamic decision making experiment. Management Science, 35(3): 321–339.
[7] Sterman JD, (1984) Instructions for running the beer distribution game (D-3679). Sloan School of Management, MIT.
[8] Lee H, Padmanabhan V, Whang S, (1997) The bullwhip effect in supply chains. Sloan Management Review, 38(3): 93–102.

12

Modelling Financial Implications of Integration Strategies

12.1 An Introductory Case Study

12.1.1 Overview

In order to migrate to higher assurance of supply levels through SCM integration, many hi-tech SC inbound business models pay special attention to financial restructuring. Financial aspects can be as important as operational aspects for supply assurance purposes. However, understanding the economic or strategic value of possible financial restructuring alternatives is sometimes not obvious to the organisations. At the same time, defining requirements and possible alternatives in order to move the organisation forward to realize that value can be a difficult task. Finally, implementing the best of possible alternatives may also be an important challenge.

The aim of this case study is to explore new designs of financial flows within a supply network competing on its capabilities and lead by a Channel Master (ChM).

In this case, the ChM had identified the following different sources of future operating profit in his inbound SC business model:

- focused, forecasting and order management, resource planning and supplier contracting;
- improved intelligence and efficiency in logistics;
- *increased cash-to-cash efficiency in inbound chain;*
- increased procurement volume and customer breadth;
- greater control over allocation of regional contract manufacturer (cm).

As a first step a team or work force has been assigned to study potential financial restructuring strategies and their implications in the assurance of supply and in the generation of future operating profits. The third of the above bullets "*Increased cash-to-cash efficiency in inbound chain*" needed to be developed.

12.1.2 Understanding Financial Problems in Contract Manufacturers

The modelling team had been developing several high level models to understand financial inefficiencies found in the financial chain. A first step was to draw a graph with the traditional supply chain cash-flows as in Figure 12.1.

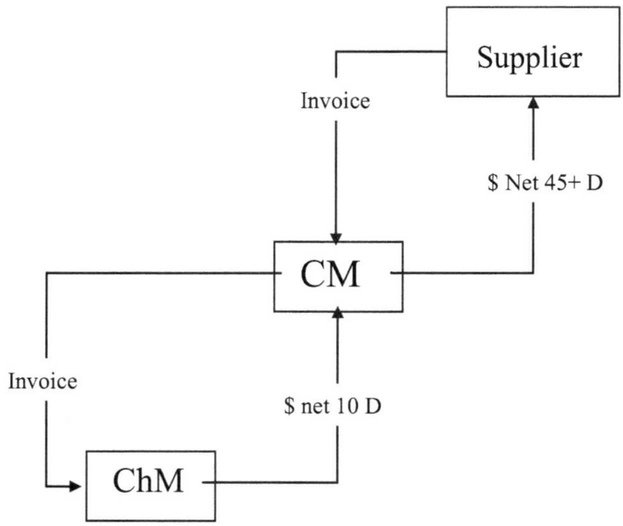

Figure 12.1. Traditional financial chain

In the graph in Figure 12.1 it is possible to appreciate how:

- CM "float" is financed by ChM and suppliers;
- supplier cash flow is variable;
- supplier's have credit high risk exposure;
- supplier ar turns are low;
- there are high inbound transaction costs.

After this first step, simple cause-effect diagrams such as that in Figure 12.2 were discussed to understand the topic and how assurance of supply could be jeopardized when product demand increases due to cash management dynamics found in contract manufacturers. Pressures to reduce ChM costs besides lower CM Payables/Receivables ratio (produced by a sudden demand increase) could reduce CM cash, producing longer periods of sales outstanding (days of sales outstanding — dso) increasing supplier working capital requirements to a point that service levels of the supplier could suddenly collapse.

The team decided to explore different options to restructure inbound financial dynamics and the following possibilities were discussed:

- Option 1: restructuring flows (1);
- Option 2: restructuring flows (2);
- Option 3: adding a "capital function".

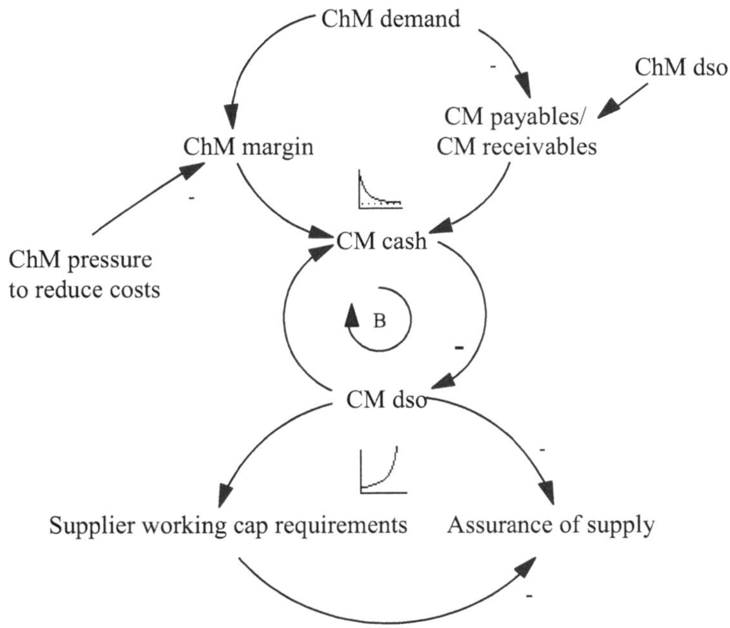

Figure 12.2. Structural inefficiencies found in the financial chain

12.1.3 Defining New Schemes

A first option would be to restructure the financial flows as presented in the graph of Figure 12.3. That option would represent the following changes:

- ChM takes 15+ Days of float;
- logistics or netting improvements increase ChM float;
- ChM payment to supplier discounted for credit risk (7–10%) and handling costs (0.25–0.50 %);
- improved cash flow and lower risk to supplier.

The second option would be to restructure the financial flows as presented in the graph of Figure 12.3. but now with a different implementation as follows:

- ChM material costs would be discounted 3–5% for assuming supplier risk and money cost;
- supplier AR turns to increase about 40% and cash flow would be stabilized;
- lower CM inventory increases WC, compensating for lost float.

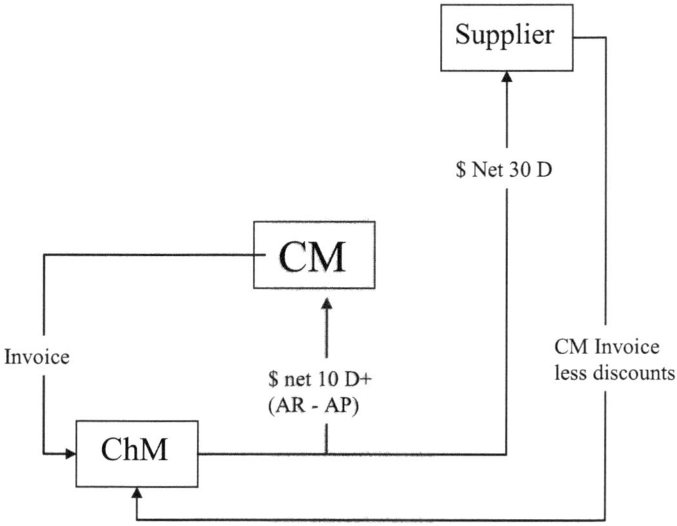

Figure 12.3. Financial chain restructured according to option 1

Finally, a third option would add a capital function to the financial chain, as presented in the graph of Figure 12.4. that would also transform the financial flows as depicted in the same Figure.

Team estimations were that, under this scenario, the following considerations would come into play:

- ChM would take 15+ Days of float;
- logistics or netting improvements would increase ChM float;
- ChM payments to supplier discounted for credit risk (7–10%) and handling costs (0.25–0.5%);
- improved cash flow and lower risk to supplier.

The team expectations regarding the advantages or disadvantages of this last option *vs* previous ones were presented in Table 12.1.

At this point, the team realised that a tool to experiment on these schemes in scenarios of variable demand was becoming a must. The need for dynamic modelling tools was clearly appreciated and a modelling project was launched at this point.

In the following section of this chapter we will present a model which allows for exploring these and other financial risk management alternatives. The aim of the modelling work is to show the reader the way to analyse financial limitations when managing supply chains that may arise for many different reasons.

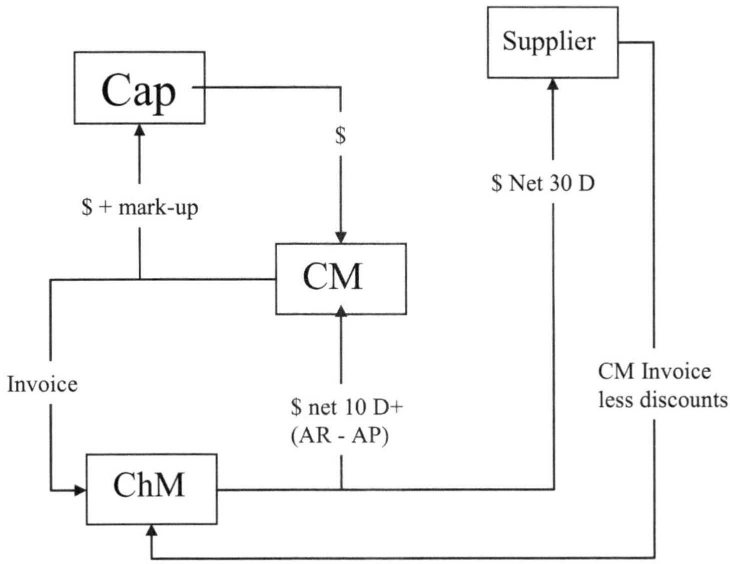

Figure 12.4. Adding a capital function (Cap) scenario

Table 12.1. Results for backlog of the node (ordered units)

	ChM	CM	Supplier
Advantages	- Improved operating margins - Greater inventory flexibility - Little asset exposure (high ROA) - Less threat of supply interruption - Increased negotiating opportunity on materials	- Lower inventory – higher ROA - Lower financial transaction costs - Greatly reduced work cap required - More competitive – improved ROA	- Improved ROA (40% improvement to AR turns) - Lower credit exposure (risk) - Improved cash flow (assurance) - Less working capital (inventory)
Disadvantages	- Some added OH - Increased financial commitment's through the definition of a new actor - Model is easy to copy	- Loss of float - Decreased negotiating leverage with suppliers - Increased financing costs (loss of float)	- Lower margins

12.2 Modelling Materials, Information and Financial Flows

12.2.1 SC Financial Variables

In order to proceed with the model development and discussion, we will first describe the notations and definition of the main variables as follows (please note that the explanation for each of these variables will be given later in the chapter).

Financial variables:

C_t^i	Cash of the node i in time t
IV_t^i	Inventories value of the node i in time t, includes materials in the pipeline plus those in the finished inventory
R_t^i	Node i accounts receivable in time t
Py_t^i	Node i accounts payable in time t
Cos_t^i	Cost of sales of node i in the period t
Cpm_t^i	Cost of purchased material of node i in the period t
Cps_t^i	Cost of production/shipping of node i in the period t
Sr_t^i	Sales revenue of node i in the period t
Sc_t^i	Sales collections of node i in the period t
Mpu_t^i	Materials purchases of node i in the period t
Mpy_t^i	Materials payments of node i in the period t
Cf_t^i	Cash flow of node i in the period t
Iwc_t^i	Increases of working capital in node i in the period t
Fe_t^i	Financial expenses in node i in the period t
Csf_t^i	Cumulative cash flow of node i in time t
Ab_t^i	Available bank credit of node i in time t
Mor_t^i	Maximum orders to place by node i in time t
Icr^i	Cash requirements per unit of material flow in node i

Financial parameters:

Cm^i	Unit contribution margin of node i
Pm^i	Price of a unit of product shipped from node i in time t
Wso^i	Weeks of sales outstanding of node i
Mb_t^i	Maximum bank credit of node i in time t

12.2.2 Considerations About Financial Statements

One of the most important responsibilities of the treasurers of different nodes of the supply chain is the management of the sources and use of funds. While making sure that cash is available to meet short-term needs, such as payrolls and invoice payments to the other nodes, treasurers must plan for strategic funds management to facilitate long-term growth via capital expansion or acquisition.

The tool for this kind of analysis is the "sources and use of funds statement" that may be estimated for any interval of time. The change in the SC node's cash position will be defined as the difference between sources and uses of funds (the reader is referred to Weston and Copeland – see pp 21–25 in [1] – for the implications of the elements of this financial statement). Another possibility would be to define increases in cash balances as a use of funds and decreases as a source. Then total sources would have to equal total uses. Normally, sources and uses statements adopt this procedure.

Since there is a multiplicity of factors impacting a firm's cash position, we pay special attention to those aspects that are related to the income from operations, and to the increments of the net working capital (an overview of this financial model is shown in Figure 12.5 where for sake of simplifying the presentation, we have not included depreciation and other non-stationary costs, thus making the inflow equal to its current income). In this context, the importance of inventory will be analysed for the overall financial picture. Note that inventory is frequently the largest asset in the SC and source of controllable costs [2].

In our analysis below, we make the following assumptions regarding the statement of changes in the financial position, and for our simulation horizon that does not exceed 1 year (52 weeks):

- Regarding the uses of funds:
  - there is no gross fixed assets expansion;
  - there is no dividend assigned to stockholder.
- Regarding the sources of funds:
  - there is no increase in long term debt;
  - there is no new equity offering during the time of the analysis;
  - there is no net fixed assets reduction;
  - there is no credit regarding production/shipping costs.
- Regarding the cash generation:
  - product contribution margin is the same within the product volume ranges of the simulation.

12.2.3 Modelling Financial Flows

Taking into account these considerations, the change in the SC node's cash position is defined as in Equation (12.1) below:

$$C_t^i = C_{t-1}^i + Ci_t^i - Iwc_t^i \tag{12.1}$$

where

$$Ci_t^i = Sr_t^i Cm_t^i - Fe_t^i \tag{12.2}$$

$$Iwc_t^i = IV_t^i - IV_{t-1}^i + R_t^i - R_{t-1}^i + Py_{t-1}^i - Py_t^i \tag{12.3}$$

Equation 12.1 implies that sources of funds will be those obtained from operations (current income), while uses of funds will be increases in the net working capital. This means that an increase in inventories is a use of funds because some product has to be bought. Further, account payables increase the available funds because, in effect, the node has borrowed from suppliers (see Figure 12.5).

Variables in the right hand side of Equations 12.4, 12.5 and 12.6 are basically "co-flows" of the material ones, defining the levels of inventory value, receivables and payables of the node. The cumulative income in Equation 12.7 is used as a metric of the model to assess system's performance:

$$IV_t^i = IV_{t-1}^i + Cpm_t^i + Cps_t^i - Cos_t^i \tag{12.4}$$

$$R_t^i = R_{t-1}^i + Sr_t^i - Sc_t^i \tag{12.5}$$

$$Py_t^i = Py_{t-1}^i + Mpu_t^i - Mpy_t^i \tag{12.6}$$

We could only study just the policies related to reductions/increases in the credit period between the nodes (that could decrease/increase the delay between the time of a sale and the cash flow from that sale, but also lower/rise the volume of unit sales). However, to be complete, we will also consider that these aspects are established within the existing contractual terms, in the agreements between the supply chain nodes (to study such a firm's dynamics we refer the reader to Lineys [3]):

$$Cci_t^i = \sum_{k=0}^{k=t} Ci_k^i \tag{12.7}$$

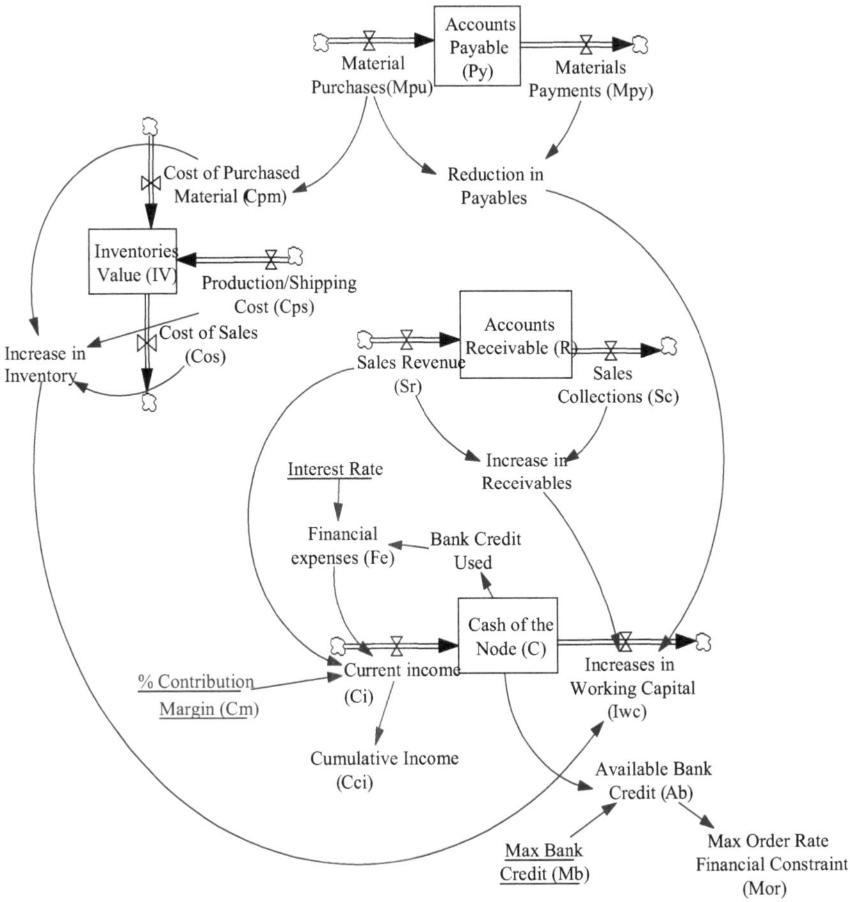

Figure 12.5. Overview of the financial model through a "stock and flow" diagram, and for a generic node of the supply chain (taken from [4])

12.3 Integration with Financial Limitations

Let us now assume that each SC node has an established price and credit with its SC partners. Suppose that a node could be exposed to a financial limit (for instance, there is a limited bank credit available for a certain period of time for that particular node). Furthermore, suppose that demand for the node's products increases and it requires a consumption of cash (in net working capital) which is higher than the cash generated by the node operations. Clearly that node could experience financial constraints impacting its operations. How much could the financial constraint of a node impact its current income and the global SC income? Would this impact depend on the integration of the SC? In order to answer these questions, we have to ascertain the node's reaction to the constraint. If we assume that the node cannot delay payments to suppliers, cannot buy cheaper parts, cannot get early payments from customers, and cannot conveniently increase the price of the products sold, the only possibility would be to order less from the suppliers. This would decrease the service level to the customers since the node will be holding less safety stock.

$$Ab_t^i = MAX\ (Mb_t^i + C_t^i, 0) \tag{12.8}$$

$$Icr_t^i = L_t^i\left[\frac{\left(Pm^{i-1} + Pm^i\left(1 - Cm^i\right)\right)}{2}\right] + ss^i Pm^i + (Wso^i - Cm^i)Pm^i - Wso^{i-1}Pm^{i-1} \tag{12.9}$$

$$Mor_t^i = I_{t-1}^i + \frac{Ab_{t-1}^i}{Icr^i} \tag{12.10}$$

Equations 12.8, 12.9 and 12.10 model the process that the members would follow to decrease their purchases to suppliers when they are limited by financial constraints. If increasing their purchase rate could lead to higher cash utilisation, they would estimate the maximum affordable increase in purchase rate. This is done by dividing their current available bank credit by the cash requirements to increase a unit of their materials flow.

Equation 12.9 estimates the cash requirements by obtaining the marginal consumption of cash produced if the flow of orders and materials increases by one unit along the SC. Each node would therefore need to fund more work-in-process, safety stock, and customers credit A/R and would receive some funds from the income of the increase in sales and from the suppliers credit A/P.

The maximum order rate to the supplier could now be obtained by adding the results of a previous division to the current build rate and pipeline input to the node as shown in Equation 12.10.

Finally, new orders to be placed from the suppliers shown in Equation 12.11 would be the minimum between the order rate formulated for each integration level of the SC (see Equations 11.8 and 11.11), and the maximum value obtained in Equation 12.12. New Equation 12.11 will also change Equations 11.1 and 11.2 to Equations 12.12 and 12.13 as follows:

$$DC_t^{i+1} = MIN\ (D_t^i, Mor_t^i) \tag{12.11}$$

$$S_t^i = \begin{cases} B_{t-1}^i + DC_t^{i+1}, & \text{if } Y_t^i \geq B_{t-1}^i + DC_t^{i+1} \\ Y_t^i, & \text{if } Y_t^i < B_{t-1}^i + DC_t^{i+1} \end{cases} \tag{12.12}$$

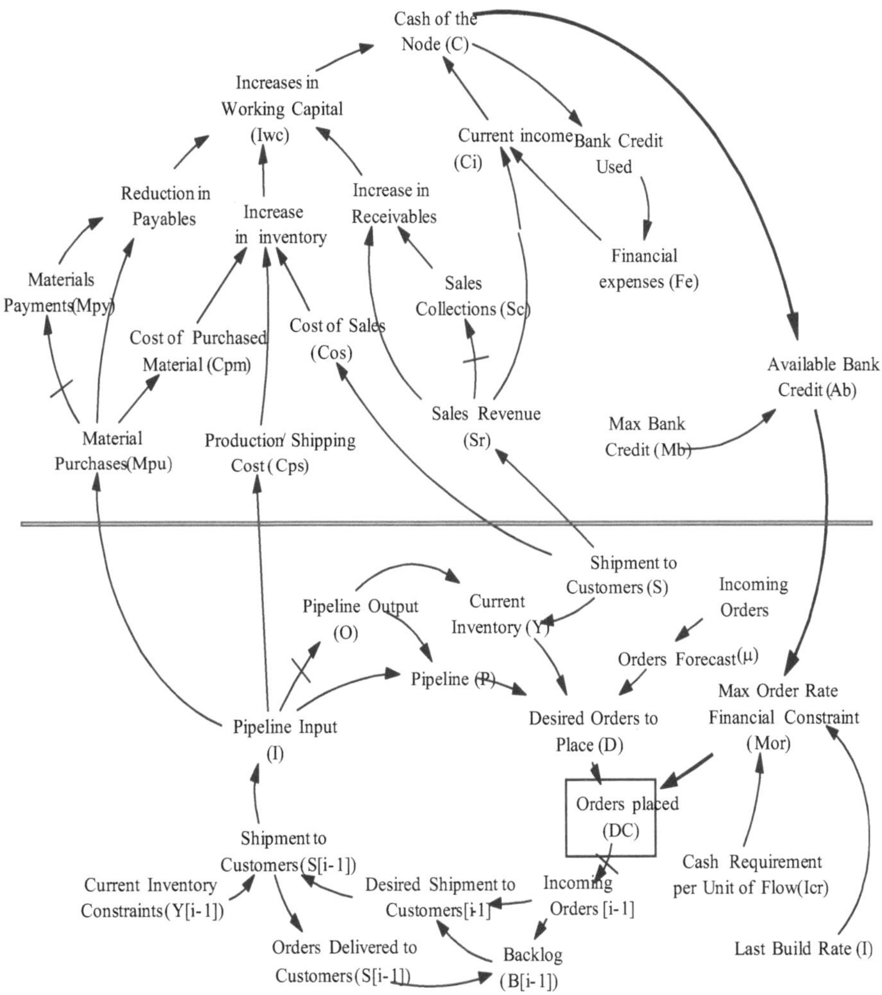

Figure 12.6. Influence diagram showing the interface between financial and material flow variables for a node of [*i*] (default variables) of the supply chain (taken from [4])

$$B_t^i = B_{t-1}^i + DC_t^{i+1} - S_t^i \tag{12.13}$$

Figure 12.6 shows the relationship between the material, information, and financial flow variables for a node of the supply chain, as formulated in Equations 12.8 through 12.13. Figure 12.6 describes how the financial constraints may appear in the node i, limiting the orders to be placed to node i–1, especially at times when demand of node i+1 ("incoming orders" in the figure) could be growing (see an example of the behaviour of these variables, for PIBSC and node "Factory", in

Figure 12.7, where it is assumed that a financial constraint in factory *i* will then limit the possibility to place orders to their suppliers). Note how, by default, some of the variables of node *i* of the SC correspond to the previous node *i*-1 ([i–1]). For example, orders placed by node *i* (DC in Figure 12.6) will increase the backlog of node *i*-1 (B[i–1]). Also some of the variables are redundant and not defined in our mathematical model (Inventory investments, Reduction in Payables, Increase in Receivables or Bank Credit Used), but are added to facilitate the interpretation of the formal model.

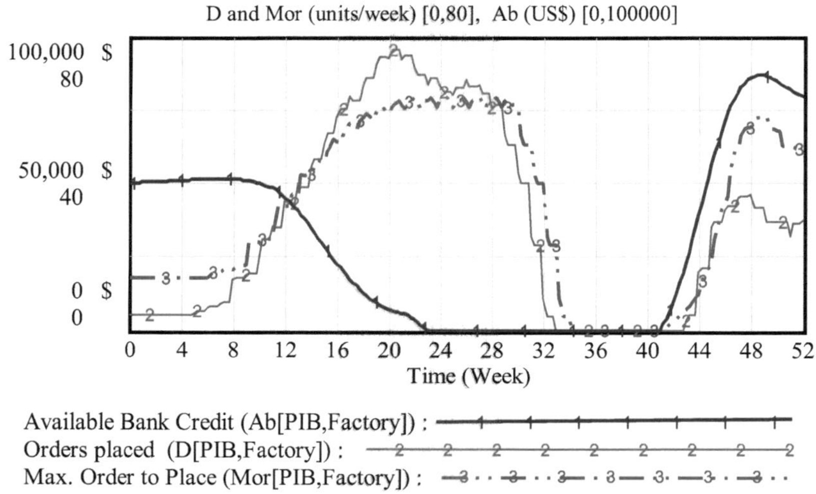

Figure 12.7. Graph showing Maximum order rate (Mor), *vs* Orders Placed (D) according to financial constraints limiting the orders to place for a PIB SC

12.4 Results with No Financial Limitations

These results are produced for the example in Section 11.6 of the previous chapter. Table 12.2 shows values for the variable "cash of the node" (C). This variable has been initialised to zero in this simulation study. In the event that C reaches a negative value, the bank credit is used, and financial expenses are paid. In this first scenario we assume that there is no limit to the money borrowed from the bank.

12.5 Integration with Financial Limitations for All Nodes

As shown in Table 12.2, different nodes have different cash requirements under different stages of the integration process. The idea of the second simulation study is to assess the impact of financial constraints of equal magnitude for all nodes in the whole SC. To avoid excessive output data, we have selected just one operational metric (mean total inventory (*P*+*Y*)) and one financial metric (cumulative income (*Cci*)). At the same time, and for the same purpose, we have selected just the first and the last nodes (factory and retailer) to present the data.

Table 12.2. Results for cash of the node (US$)

Variable **Cash of the Node (C)**	Min	Max	Mean	StDev
(NI, Factory)	–12,415	5,627	–5,068	5,936
(NI, Distributor)	–38,359	22,792	6,397	8,138
(NI, Wholesaler)	–58,719	22,056	4,894	16,104
(NI, Retailer)	–92,020	21,532	–5,787	29,154
(PIA, Factory)	–20,241	46,963	5,749	19,550
(PIA, Distributor)	–19,724	56,053	12,828	18,801
(PIA, Wholesaler)	–37,987	41,547	5,261	15,876
(PIA, Retailer)	–51,863	53,377	–2,594	25,039
(PIB, Factory)	–83,496	37,615	–32,445	37,112
(PIB, Distributor)	–45,152	43,888	–2,352	24,697
(PIB, Wholesaler)	–14,253	57,440	11,082	17,891
(PIB, Retailer)	–52,949	52,401	–2,946	26,253
(FI, Factory)	–44,248	31,611	–1,635	21,620
(FI, Distributor)	–24,013	49,993	6,850	20,668
(FI, Wholesaler)	–29,606	51,324	5,140	21,232
(FI, Retailer)	–51,195	59,127	–496	25,430

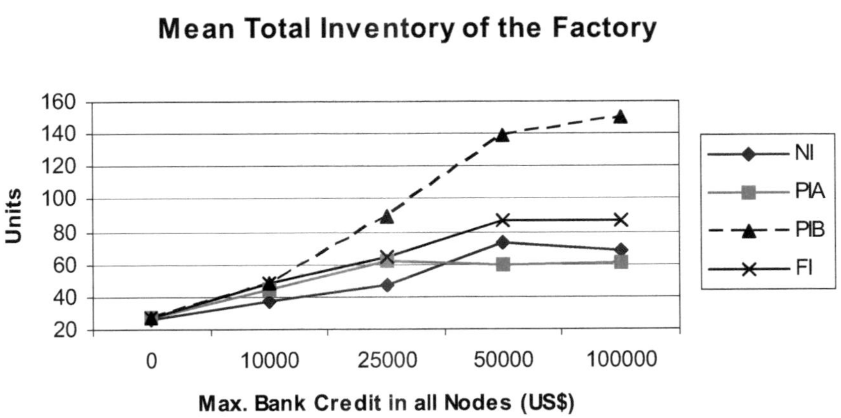

Figure 12.8. Mean total inventory (P+Y) of the Factory (units)

Table 12.3 shows total funds generated by operations along the simulation by different nodes. FISC seems to perform better, especially for the retailer, followed by PIBSC, PIASC and NISC.

Table 12.3. Results for cumulative income of the nodes (US$)

Variable **Cumulative Income (Cci)**	Min	Max	Mean
(NI, Factory)	0	93,239	28,831
(NI, Distributor)	0	91,072	31,033
(NI, Wholesaler)	0	86,009	31,805
(NI, Retailer)	0	112,774	39,271
(PIA, Factory)	0	90,792	36,402
(PIA, Distributor)	0	93,939	39,992
(PIA, Wholesaler)	0	103,393	41,486
(PIA, Retailer)	0	135,770	51,829
(PIB, Factory)	0	79,821	36,081
(PIB, Distributor)	0	94,428	39,392
(PIB, Wholesaler)	0	105,020	42,287
(PIB, Retailer)	0	137,437	52,606
(FI, Factory)	0	88,753	40,067
(FI, Distributor)	0	100,324	43,571
(FI, Wholesaler)	0	108,868	45,155
(FI, Retailer)	0	144,083	57,009

We show and briefly discuss the results for five scenarios according to the maximum bank credit (Mb) which is available for all the nodes: US $ 0, 10,000, 25,000, 50,000, and 100,000.

These results show how inventories and current incomes could change with the change in the sequence of integration and the existing financial constraints (see Figures 12.8 to 12.11). For instance, PIBSCs produce higher factory inventory levels (Figure 12.8) and lower incomes (Figure 12.10) than the SC structures with no financial constraint (worse performance structure).

However, note how PIBSC performs better than the NISC, and similar to or better than the FISC and PIASC when financial constraints increase.

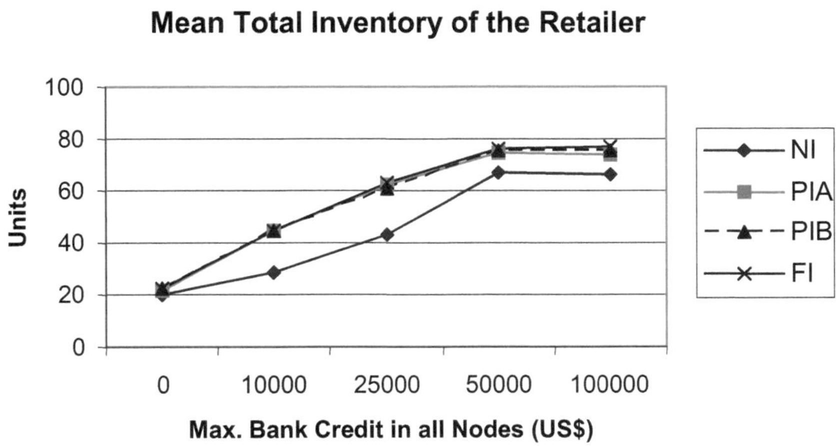

Figure 12.9. Mean total inventory (Y+P) of the retailer (units)

Figures 12.10 and 12.11 show how PIASC and FISC maintain higher levels of cumulative income when constraints increase. Figures 12.9 and 12.11 show that any integration structure, partial or full, always produces a better performance for the retailer.

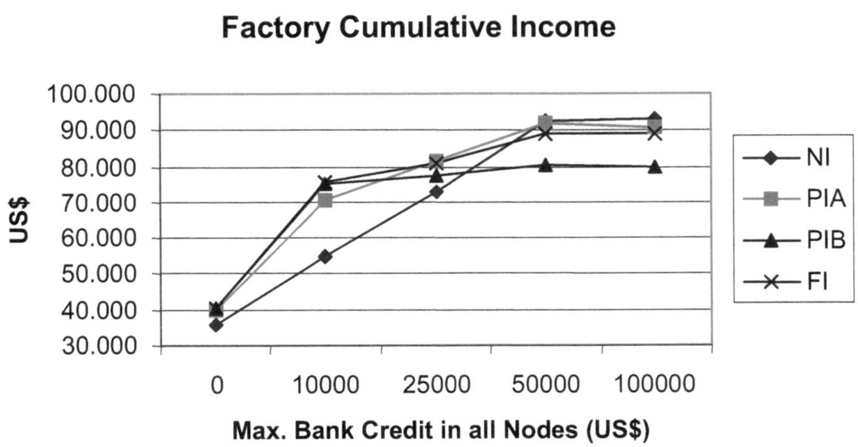

Figure 12.10. Factory Cumulative Income (Cci) (US$)

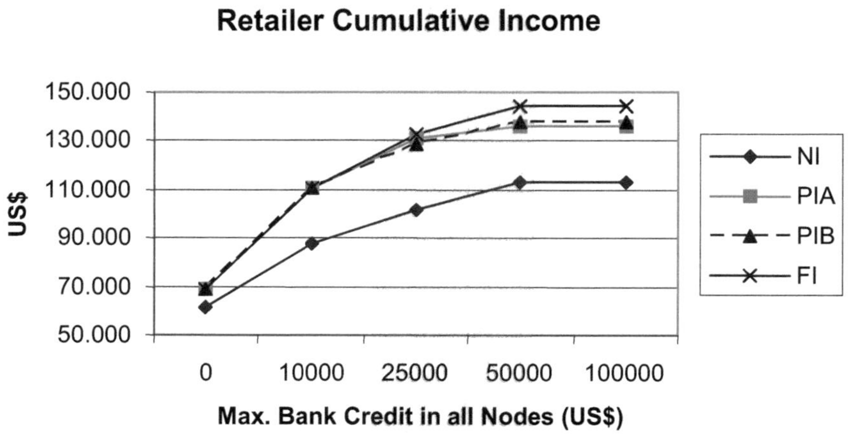

Figure 12.11. Retailer Cumulative Income (Cci) (US$)

The above results show how collaborative planning without any previous collaboration and discussion to generate a common forecast (i.e. PIBSC structure) could lead to inefficient supply chain performance when there is no financial constraint. For the same reason, releasing financial constraints through electronic payments tools in a PIBSC could not necessary be beneficial for the SC performance, and it seems reasonable, according to these results, to apply these tools once the common forecasting is in place.

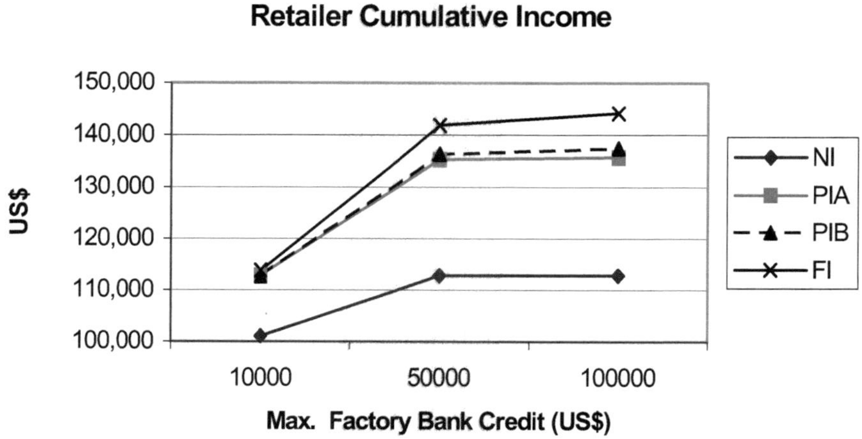

Figure 12.12. Retailer Cumulative Income (Cci) (US$) for Factory Financial Constraints Only

12.6 Financial Limitations at a Single Node

The model presented allows this interesting analysis. Many SC engineers wonder, for instance, how a supplier's financial situation can impact the whole SC performance. This problem can be explored with the model. It can be explored increasing the factory financial constraints only, and then obtaining the impact on the retailer's cumulative income.

Results are presented in Figure 12.12, where the systemic nature of the SC can be observed. The retailer's cumulative income drastically decreases due to the factory's financial constraints. In fact, the income amounts are very close to the amounts that we would observe when all nodes were constrained (see Figure 12.13 for FI results).

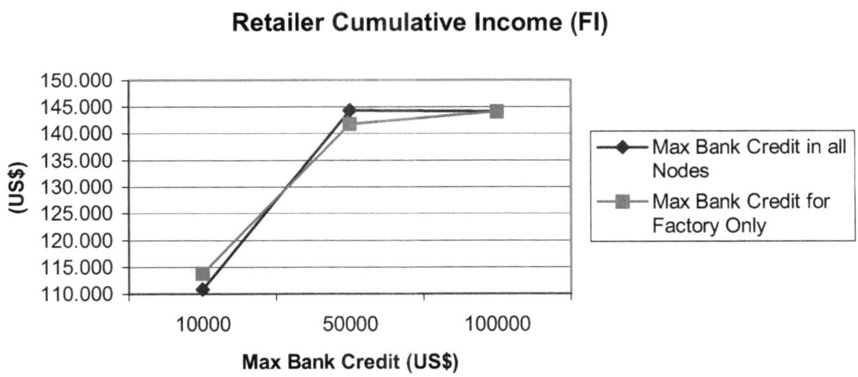

Figure 12.13. Retailer FI Cumulative Income (Cci) (US$) comparison for all Nodes *vs* only factory financial constraints

12.7 Concluding Remarks

In this chapter a comprehensive model to study the operational and financial benefits of using various e-collaboration tools has been developed.

Our computational results also show that it is risky to install e-collaboration tools for electronic payment when collaborative forecasting is not in place in the SC. Decreases in financial constraints could lead to an unnecessary increase in inventories without improving SC performance. Local financial constraints can heavily impact the operational and financial performance of the entire supply chain. At times this impact could be very close to that produced by a global financial constraint at all the nodes of the SC. Therefore, helping the weakest financial node of the chain should be a main concern of the SC engineers and analysts, and not treated as an SC local issue.

12.8 References

[1] Weston JF, Copeland TE, (1989) Managerial finance (Eight Edition). Chicago: The Dryden Press.
[2] Rockhold S, Lee H, Hall R, (1998) Strategic alignment of a global supply chain for business success. POMS Series in Technology and Operations Management. Vol.1, Global Supply Chain and Technology Management, 16–29.
[3] Lyneis JM, (1980) Corporate planning and policy design: a System Dynamics approach. Cambridge, MA: MIT Press.
[4] Crespo Marquez A, Bianchi C, Gupta JND, (2004) Operational and financial effectiveness of e-collaboration tools in supply chain integration. European Journal of Operational Research, 159(2): 348–363.

13

Exploring the Use of Manufacturing Control Techniques in Virtual SC

13.1 Virtual Manufacturing in Modern Supply Chains. Comparing SC Integration Levels to Push-Pull Manufacturing Schemes

In the literature we can find two major manufacturing control strategies, named push and pull manufacturing. Push manufacturing is characterised by manufacturing to forecast, and emphasises batch processing and lot sizes. Each area runs at maximum capacity, and the material is pushed downstream. Push planning methods include MRP, reorder points, and optimum order quantities.

Pull manufacturing is a visual replenishment of goods, where only what is demanded is produced by the next work centre. The most important pull planning methods are based on visual signals or on cards or tags. Pull planning methods include the determination of the number of cards (Kanbans, setting the maximum WIP levels) and inventory review periods. Stock based systems and CONWIP (Constant Work In Process) systems can also be considered as pull or long pull systems.

The main difference between push and pull control schemes is the effect that changes in demand, produces on the production line materials. It was this effect that originated this "push-pull" terminology as a name for these manufacturing materials control systems.

Current virtual supply chains operations show certain analogies to previously mentioned material control models. During the last few decades many companies have introduced enterprise resource planning systems (ERP systems) in the search for operational effectiveness and profitability. These systems are inward-looking and focussed on efficiency [26]. They use advanced statistical and operational research tools *to push* products efficiently to the customers trying to match forecasted market demand. Technology-enabled supply chain systems allowed vertical process integration producing better utilisation of fixed assets and working capital. An example of a firm following this model is that of a car manufacturer.

The investments in property, plant and equipment of these companies requires optimal operations in order to match predicted car demand.

On the other hand, recent Internet technologies have enabled information systems to interact with one another. Business networks have emerged within and outside the companies to satisfy the needs for competitive flexibility. As opposed to the previously explained push model (*market economy model*, as named by Reddy and Reddy [26]), where firms push products to the customers based on forecast, this new *network economy model* (pull model) is geared toward the customer pulling customised products to satisfy his or her individual needs. In the ideal pull model, every shipped product is built to order per customer specifications. The order fulfillment process is visible to all trading partners and they interact directly with this fulfillment process. All trading partners collaborate to provide a single face to the customer. Business networks — internal to the firm and across the supply chain — would utilise Intenet technologies to present a single face to the customer.

13.2 Hybrid Push-pull Manufacturing Schemes Used for SCM

As we have previously mentioned, some authors have developed important advances in order to integrate and improve the performance of the supply chain (SC) by sharing information through communication technologies with a centralised and/or decentralised approach. The concepts of push and pull represent two extremes of a continuum. Normally, depending on the type of product and the location of a firm's product in the product life cycle, the firm may prefer one form over the other. In reality most firms may incorporate elements of both models. For instance, during the growth stage, product availability is paramount to the firm. A pull model with a collaborative infrastructure can support the marketshare objectives for the firm during this period. Over time, however, as more competitors enter the market segment, eroding margins may force the firm to optimise the utilisation of its productive resources. The focus may shift from collaboration to synchronisation of the supply chain [26], from the pull to the push model.

In the following section we will explore how to use, in a virtual supply chain, a long-pull manufacturing technique such as CONWIP (CONstant Work In Process). The idea is that CONWIP may now control materials flowing through the entire supply chain. Of course, a virtual centralised supply chain control system is required to be able to apply this long pull manufacturing control technique for the supply chain control.

13.3 Sample CONWIP Driven Virtual Supply Chain

13.3.1 Introduction to the Case Study

An alternative to improve the performance in an SC may be to introduce a CONWIP Supply Chain (CONWIP SC) policy with a centralised approach in the supply chain management (SCM) [27]. In this section we review the literature to present the benefits of the CONWIP system in different production environments and discuss the possible utilisation of this system to manage the entire supply chain. A Characterisation of the CONWIP-based approach to supply chain management is presented, and a simulation model to explore and evaluate the advantages of this strategy, in comparison with a Fully-Integrated Supply Chain (FI SC) is developed. We continue this analysis for different levels of demand volatility and material flow constraints along the chain, and we optimise parameters to ensure the best performance of each control supply chain policy in each scenario. This research shows that CONWIP SC policy may offer advantages and improved performances in global metrics while both policies offer similar service levels. The main advantages are a smaller average of orders placed, less impact of demand variability on the ordering policy, shorter average finished goods inventory and work in process (FGI+WIP) levels and potential inventory cost, substantial comparative funds savings to run the business model, and easier control of inventories.

13.3.2 The CONWIP SC Approach

CONWIP Supply Chain is an approach by which we attempt to improve the Supply Chain performance through an extension of the closed production control system CONWIP–CONstant Work In Process [1].

CONWIP is a "long-pull" production technique, generalised from a kanban system. In CONWIP systems cards are assigned to the whole production line [1]. When starting the production, all available cards are located at the beginning of the line (on a bulletin board). When orders arrive, and there are enough available cards in the system, the necessary cards are attached to the order, and together they proceed through the production line. When the order is processed completely in the line, and leaves the final station, the card is dropped off and released back to the beginning of the line. No order can enter the line without its corresponding card, *i.e.* if orders arrive and no free card exists, the orders accumulate as backorders, from where they will be discharged as cards are released. Intermediate buffers are established between two consecutive stations, driven by an FCFS (first-come first-served) discipline. The detailed flow control mechanism of CONWIP is extensively discussed by [2].

This is the mechanism by means of which the WIP stays constant (whenever the demand is above the capacity). To design a pattern of the CONWIP system, two fundamental questions should be analysed, the administration of the backlog and the computing of the number of cards. Keeping in mind the time when information about demand at the final buffer is forwarded and the path the information takes, Figure 13.1 shows the operative difference between pull and CONWIP control systems.

CONWIP SC is defined in this book as a production-distribution system, in which the production line of each firm has a similarity to a "work center" being a part of a "global line" of supply. The set of cards mentioned in the description of CONWIP system, extends now to a virtual center of control that governs the supply chain and manages the parts flow and the inventories along the chain. When orders arrive at the final node, the production orders and required materials are released to the first node taking into consideration its production capacity constraints. There is a unique and centralised control of the backorders of the SC. Thus, the centralised information control through Internet type of tools is critical in this context.

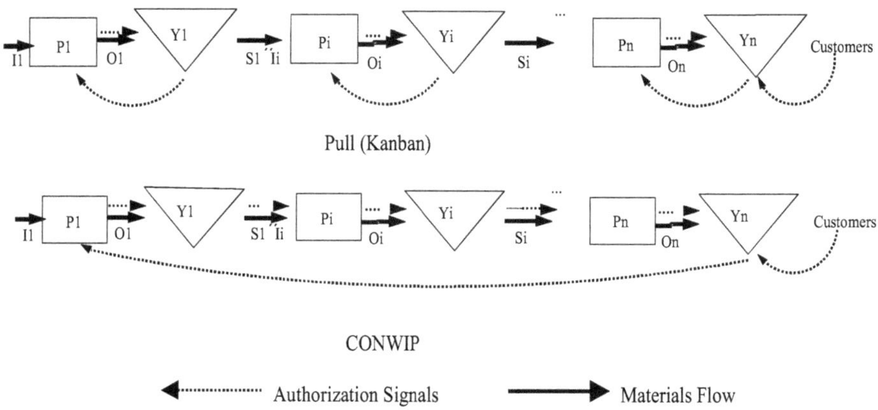

Figure 13.1. Pull (kanban) *vs* CONWIP control schemes

In Section 13.3.3 there is a review of the CONWIP system and then a discussion of the CONWIP SC is presented. In Section 13.3.4, we characterise a CONWIP SC model and compare it with a model for an FI SC. In Section 13.3.6, CONWIP SC simulation model outputs are discussed to validate the main variables behaviour. In Section 13.3.7, certain metrics are selected and their results are obtained and discussed to evaluate and compare CONWIP SC *vs* FI SC policy. Finally we present conclusions of the case study in Section 13.3.8.

13.3.3 CONWIP in a Production System *vs* CONWIP in a SC

In the following paragraphs we will present a review concerning the most important studies carried out up to the present day, showing the advantages of CONWIP control systems over other control systems.

A review of relevant literature shows how the CONWIP system possesses some superior characteristics over the other pull systems. Amongst others, we have decided to highlight the following:

- It is simpler in the sense that only a single card count setting is required instead of a card count for each workstation [2].
- It can accommodate a changing part mix, due to its use of line-specific cards and a work backlog [2].
- It can accommodate a floating (mix-dependent) bottleneck, due to the natural tendency of WIP to accumulate in front of the slowest machine [2].

- It introduces less operator stress due to a more flexible pacing protocol [2].
- In a flow line that produces a single part type, Spearman and Zazanis [4] have shown that CONWIP produces a higher mean throughput than Kanban. In the same scenario, Muckstadt and Tayur [15, 16] showed that CONWIP produces a less variable throughput and a lower maximal inventory than Kanban.
- Although some researchers have shown that a kanban system would have lower WIP levels than a CONWIP system with the same throughput under some certain conditions [7], most researchers have pointed out that CONWIP would result in lower WIP levels than kanban system with the same throughput in most cases [1, 4]. In other words, a CONWIP system yields larger throughput than a kanban system for the same number of containers (maximal inventory) [8, 9], even for systems with yield losses [5]. This is thanks to the existing global control (of orders, backorders, and inventory) in a CONWIP system.
- It copes with flow shop operations with large set-up times and permits a large product mix [4].
- According to the simulation study carried out by Roderick *et al.* [10], a CONWIP system is superior to other production control systems with respect to due dates and cycle times.
- Bonvik and Gershwin [11] found that a simple CONWIP control policy outperforms kanban with respect to average inventory levels when subject to the same requirements on throughput and service level (fill rate). They also concluded that, when the system operates close to capacity, the hybrid control combining CONWIP and kanban improves the inventory levels further.
- Huang *et al.* [12] found that the CONWIP production control system is very efficient for the production and inventories control of semi-continuous manufacturing. According to the authors it can greatly reduce the WIP, decrease the average inventory and average inventory costs, and guarantee a higher throughput rate and facility utilisation.

In comparison with push systems:

- Herer and Masin [13], built a model in which they highlighted the main advantages of CONWIP over MRP systems. They affirm that the difference between the MRP and CONWIP production control systems lies in the way inventory is handled, because in MRP-ruled manufacturing systems, the amount of inventory managed in the system is theoretically unlimited (see [14] p 34). This difference results in long lead times of MRP systems, poor service levels and large work in process (WIP) and finished good inventories [15].
- Huang *et al.* [12] assure that CONWIP is superior to push systems when the production system runs under the highest possible throughput rates, and CONWIP also appears to alleviate a problem found in many push systems called "overtime vicious circle". This concept is related to real life steady state cycles in a plant, as a consequence of capacity computation,

randomness in job arrivals creating the bottleneck process to starve, and later increasing WIP and cycle time, which leads normally to the "one time" authorisation of overtime until WIP and cycle time decreases again (see [2], pp 284–285).
- Hopp and Spearman [2] add that the WIP level is directly observable, while the release rate in a push system must be set with respect to (unobservable) capacity, it requires less WIP on average to attain the same throughput, it is more robust to errors in control parameters and it facilities working ahead of a schedule when favorable circumstances permit it.

CONWIP, however, may have its own disadvantages:

- CONWIP does not always generate the smallest total number of trips between stages [16].
- CONWIP may also require a larger storage space between alternate stages than kanban type flow lines, because all full containers (with attached withdrawal cards) may accumulate between any pair of alternate stages [16].
- Graves *et al.* [17], assures that a serious drawback of CONWIP is that it does not consider the impact that a bottleneck work center may have on the performance of a manufacturing line.

If we now concentrate on SCM policies, according to the different level of SC integration, the SC uses several push and pull elements in the movement of parts along it [3]. For instance, collaborative forecasting may introduce a push effect in the early stages of the SC, while replenishments pulses could be, at the same time, moving materials among partners, producing pull-type local effects. Since the relevant advantages of CONWIP *vs* other pull and push production systems were found, we will try to enlarge the underlying philosophy in the CONWIP production environment, searching to materialise some of those advantages in an SCM context. For instance, could we expect to maintain less WIP levels for the same fill rate of the SC when using CONWIP SC? In which scenarios would that be possible?

Compared with the fully-integrated SC (extending our literature review concerning CONWIP systems), we conclude that the following advantages could be capitalised by introducing CONWIP elements into SCM. The intermediate nodes would:

- not need to establish ordering policies, thus we may avoid the amplification of the variance of the demand signal along the chain, only a central entity would authorize the release of work on the basis of the system status, defined by the total number of cards "attached" to orders;
- not need to establish policies to fix safety inventories, reducing holding and ordering costs and inventory cash requirements, even more, using less space;
- not need to control backlog, due to centralised supply chain management;

- make simpler operative decisions (under normal conditions, in a centralised supply chain, firms would simply process and send the whole received parts);
- make easier control of materials flow and WIP, since parts are transferred between partners by means of a push effect;
- not need to carry out individual forecasts, nor do they need to know the levels of inventories of the other nodes;
- be in a position to promote teamwork, the development of products and processes, and to realise prominent cooperative agreements in order to achieve virtual financial integration.

However, in the light of existing management paradigms, managers could see disadvantages and prefix obstacles to implementing a centralised control policy (with constant work in process in the SC), for example:

- in a general sense, the SC is constituted by different firms with different goals, probably in conflict;
- information about demand would not be shared between nodes, but centralised, so that the firms would not manage individual forecasts, and ordering policy would be global, diminishing the internal control in the firms;
- limiting the total number of parts allowed into the SC at the same time, may prevent a particular firm use its full capacity;
- centralised management may impose conditions on the financial policies, as well as the productive processes and policies of the firms. Furthermore, difficulties (e.g. bottleneck, quality defects, cost overrun, etc.) in one firm, may easily affect all the chain:
- not all firms are willing to share financial and operative risks.
- CONWIP may not be well suited for the case of complex SC networks.

13.3.4 Modelling a CONWIP SC *vs* an FI SC

Before proceeding with the model development and discussion, we will first describe the notations and definition of the main variables as follows:

Material flow variables:

P_t^i	Pipeline of the node *i* (work in process inventory in the node *i*) in *t*
Y_t^i	Finished goods (or parts) inventory in the node *i* in *t*
S_t^i	Shipment rate from the node *i* to the node *i*+1 in period *t*
O_t^i	Output from the pipeline (completion rate) of node *i* in period *t*
I_t^i	Input to the pipeline (procurement rate) of the node *i* in period *t*

Information flow variables:

D_t	Incoming orders to the SC (in final node) in period *t*
B_t	Backlog in the SC in *t*
DO_t	Orders finally shipped (from last node) to end-customers in period *t*
$\hat{\mu}_t$	Forecast in period *t*
APC_t	Available production cards in *t*
OP_t	Orders placed by the SC (by first node) in period *t*
PB_t	Production backlog, *i.e.* orders which should, but cannot enter in production in *t*
DS_t	Desired shipments from the SC in period *t*
DPO_t	Desired production orders of the SC in period *t*
TY_t^i	Available total FGI in the node *i*, in *t*

Model parameters:

L^i	Cycle time for a unit in the pipeline to arrive to the FGI of the node *i*
MLP^i	Maximal load of units to be processed in the node *i*
α	Forecast smoothing factor
UC^i	Units per container in the node *i* (assumed same value *UC* for all nodes)
$TNPC$	Total number of production cards

In order to compare performance of the two SCM alternatives, we will model both systems according to the following procedure:

1. Formalisation of the FI SC model according to Crespo *et al.* [3], in which it is assumed that (a) partners make individual ordering decisions, implying that a decentralised control exists in the chain, (b) the orders in the supply chain are visible in real time, and processed accordingly and (c) partners gain access to additional information that they do not control, and use it in their planning process (e.g. forecasts of final nodes, on-hand inventories and pipelines of the other supply chain members, backlog, etc.).
2. Comparison of the equations for the FI SC model to those for the CONWIP SC model, indicating those which are different to their homologous in the CONWIP SC model (see Table 13.1).
3. Equations of the CONWIP SC model will be presented.

4. Behaviour patterns for some CONWIP SC model variables will be provided for their validation.
5. The simulation study will be carried out, to compare performance of both systems for a series of designed scenarios. We will make sure that, for each particular scenario in the study, both systems are operating in their best possible conditions (*i.e.* with the best possible value for their parameters measuring criteria are presented in Section 13.3.7). This will be done via a multi-parametric optimisation for each particular case.

Table 13.1. Model equations in fully-integrated SC *vs* CONWIP SC

Equation ([a]: Equation for this variable differs in CONWIP SC model; see Section 13.3.6)	**No.**	**Homologous No. Eq. in CONWIP SC**
Final demand: OP_t^{cust}		
Incoming orders:		
$D_t^{i+1} = OP_t^{i+1}$, from node i+1 to node i $D_t^{cust} = OP_t^{cust}$, to final node	(1) [a]	(13.1)
Desired shipments (in final node, backlog B_{t-1}^{n} are not included in Equation 13.2, since they are considered lost sales):		
$DS_t^i = B_{t-1}^i + D_t^{i+1}$, from node i to node $i+1$ $DS_t^n = D_t^{cust}$, from final node	(2) [a]	(13.2)
Shipment rate:		
$S_t^i = MIN(TY_t^i, DS_t^i, MLP^{i+1} / L^{i+1})$, from node i to $i+1$ $S_t^n = MIN(TY_t^n, DS_t^n)$, from final node	(3) [a]	(13.4)
Total available inventory (in each node):		

Table 13.1. (Continued)

Equation ([a]: Equation for this variable differs in CONWIP SC model; see Section 13.3.6)	**No.**	**Homologous No. Eq. in CONWIP SC**
$TY_t^i = Y_t^i + O_t^i$	(4)	(13.5)
Backlog (in each node):		
$B_t^i = B_{t-1}^i + D_t^{i+1} - DO_t^i,$	(5) [a]	(13.10)
On hand inventory (in each node):		
$Y_t^i = Y_{t-1}^i + O_t^i - S_t^i$	(6)	(13.6)
Finally delivered orders (from each node):		
$DO_t^i = S_t^i$	(7) [a]	(13.11)
Output from the WIP (completion rate in each node):		
$O_t^i = I_{t-L^i}^i$	(8)	(13.7)
Input to the WIP (procurement rate) of each node:		
$I_t^i = S_t^{i-1}$	(9)	(13.8a)
Pipeline (WIP) from node i to the next node i+1:		
$P_t^i = P_{t-1}^i + I_t^i - O_t^i$	(10)	(13.9)
Forecast: information about final customers, shared by all nodes (where n is final node):		
$\hat{\mu}_t^i = \alpha^n D_{t-1}^{cust} + (1-\alpha^n)\hat{\mu}_{t-1}^n,$ with $0 < \alpha^n \leq 1$	(11)	
This exponential smoothing forecast is widely used in SC modelling (see e.g. [18, 19]).		

Table 13.1. (Continued)

Equation ([a]: Equation for this variable differs in CONWIP SC model; see Section 13.3.6)	**No.**	**Homologous No. Eq. in CONWIP SC**
The desired production orders (from each node) are computed by means of anchoring and adjustment heuristic [20] with fractional adjustments coefficients (β_S , β_{SL}) for the FGI and the WIP respectively, subtracting the backlog of the supplier firm and not allowing negative values of the requested quantity [21]:		
$DPO_t^i = Max(\hat{\mu}_t^i + \beta_S(\hat{\mu}_t^i ss^i - Y_t^i) + \beta_{SL}(\hat{\mu}_t^i L^i - P_t^i) - B_{t-1}^{i-1} + ib_t^i, 0)$	(12) [a]	(13.3)
Full integration is improved when discounting the previous node's backlog. ss^i is safety stock, and ib_t^i is a variable expressing the inventory information provided to the node *i* from downstream partners, through the *i*nformation *b*ackbone in time *t*:		
$ib_t^i = \hat{\mu}_t^i(ss^{i+1} + L^{i+1}) - (P_t^{i+1} + Y_t^{i+1}) + ib_t^{i+1}$	(13)	
Released production orders and orders placed (by each node):		
$OP_t^i = Min(DPO_t^i, MLP^i / L^i)$	(14) [a]	(13.13)

13.3.5 CONWIP SC Equations

In Figure 13.2, the connection between the main CONWIP SC information flow variables (thin lines) and the material flow variables (thick lines) can be clearly appreciated.

Now we will explain how the equations are obtained for the CONWIP SC model.

Incoming orders

The incoming orders variable (D_t) takes the generated values for the orders placed by the final customers (OP_t^{cust});

$$D_t = OP_t^{cust} \tag{13.1}$$

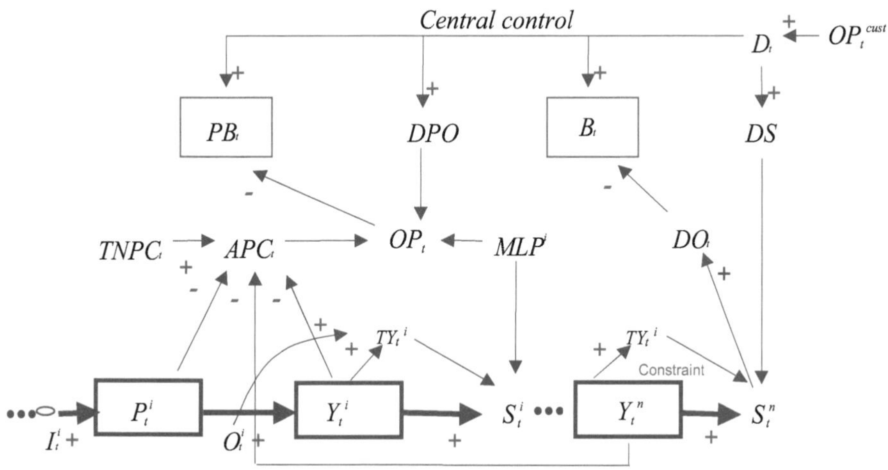

Figure 13.2. Basic influence diagram of CONWIP SC for the variables in node [i] and node [n]

Shipments

It is assumed that the incoming orders to the final node (D_t) are immediately shipped to final customers as they are received. Backlog is considered to be lost sales. Therefore, the desired shipments DS_t from the last node are equal to the orders received in each period, as follows:

$$DS_t = D_t \tag{13.2}$$

Likewise, the desired production orders to release to the first node of the supply chain, becomes equivalent to incoming orders:

$$DPO_t = D_t \tag{13.3}$$

The shipments from the node i in week t (S_t^i) depends on the available total *FGI* (TY_t^i) and the maximal load of parts to be processed in the node i+1 during its cycle time (MLP^{i+1}/L^{i+1}). When delivering finished goods, inventory constraints may appear in the node n, reducing the amount of units shipped to customers, S_t^n (see Figure 13.2). The shipment rate to final customers from final node is the minimum among the available total FGI TY_t^n and the desired shipments to final customers DS_t . The expression for the supply flow is as follows:

$$S_t^i = \begin{cases} MIN\ (TY_t^i, MLP^{i+1} / L^{i+1}) & \text{for nodes } i = 1 \ldots n\text{-}1 \\ MIN\ (TY_t^n, DS_t) & \text{for final node } i = n \end{cases} \tag{13.4}$$

For all nodes, the available total FGI is calculated as follows:

$$TY_t^i = Y_t^i + O_t^i \tag{13.5}$$

This includes the on-hand inventory in the buffer and the output from the pipeline of the same node.

Inventory, materials flow and work in process

The FGI Y_t^i in the buffer of the node i diminishes according to the units transferred to the next node i+1 or to final customers, and increases according to the rate of processed parts that arrive (initial conditions are assumed known):

$$Y_t^i = Y_{t-1}^i + O_t^i - S_t^i \tag{13.6}$$

where O_t^i is the output from the pipeline (completion rate of WIP) in the period t, calculated in the following way:

$$O_t^i = I_{t-L^i}^i \tag{13.7}$$

I_t^i is the procurement/production rate of the node i in period t, equivalent to the shipped orders by node i-1, S_t^{i-1} as follows:

$$I_t^i = S_t^{i-1} \tag{13.8a}$$

When assuming unlimited supply of parts to the first node and without variability in the market of external suppliers to the supply chain, I_t^i is equivalent to order placed by the first node OP_t, calculated by the control central system of the CS; this is

$$I_t^1 = OP_t \tag{13.8b}$$

Therefore the WIP in the pipeline is defined (assuming its initial conditions are also known) as

$$P_t^i = P_{t-1}^i + I_t^i - O_t^i \tag{13.9}$$

Backlog, procurement orders and available production cards

When the inventory constraints limit the supply of finished goods from the last node of the SC to final customers, a portion of the orders cannot be fulfilled. These orders will stay as backlog B_t (in the central control of SC), and they are considered lost sales. In contrast to FI SC, in CONWIP SC backlog is generated only in final node, from incoming orders D_t and finally served orders DO_t, in the following way (supposing initial conditions are known):

$$B_t = B_{t-1} + D_t - DO_t \tag{13.10}$$

where DO_t is the information of the final supply to end customers, calculated in the final node and centrally controlled as follows:

$$DO_t = S_t^n \tag{13.11}$$

When there are not enough available production cards to release production orders, the flow of orders to the entire supply chain is limited, and then a portion of the final customer orders cannot be released to production. Those orders will stay as production backorders PB_t (production orders not released on time), and they are computed as in the following equation:

$$PB_t = PB_{t-1} + D_t - OP_t \tag{13.12}$$

where the orders placed OP_t are obtained from the minimum value between desired production orders DPO_t, the available production CONWIP cards, converted to units and assuming the same quantity of units per container (*UC*) for all nodes, $APC_t * UC$, and the maximal production rate of first node MLP^1 / L^1, according to following expression:

$$OP_t = MAX(MIN(APC_t * UC, DPO_t, MLP^1 / L^1), 0) \tag{13.13}$$

where it is presumed that there is only one period between two consecutive orders, and it is established that orders must not be negative.

The available production cards APC_t are calculated as follows:

$$APC_t = TNPC - (\sum_{i=1}^{n} P_t^i + \sum_{i=1}^{n} Y_t^i) / UC \tag{13.14}$$

The total number of cards (*TNPC*) is calculated to maximise the throughput while the total inventories are minimised in the entire chain. This is done by using direct-search techniques applied to each particular scenario.

Financial flows and the working capital

Financial flow variables are estimated extending part of the one-stage model presented in [22] to a multiple-stage model of the supply chain. We now present the relationship between the main financial variables (related to cash requirements) for each node, using the following notation:

CR_t^i	Cash requirements (working capital) of the node i in time t
ICR_t^i	Inventory value of the node i in time t, includes cash requirements to fund the materials in the WIP plus those in the FGI
R_t^i	Node i accounts receivable in time t
Py_t^i	Node i accounts payable in time t
Pm_t^i	Price of a unit of product shipped from node i in time t
$Pwip_t^i$	Average value of a work in process unit in node i in time t
mr_t^i	Profit margin in a product in node i in time t
$dso(i)$	Weeks of sales outstanding of node i
$CumP_t^i$	Cumulative profit of node i, in time t

For the purpose of this chapter, the main variables relationships are as follows:

$$CR_t^i = ICR_t^i + R_t^i - Py_t^i - CumP_t^i \tag{13.15}$$

$$Py_t^i = Py_{t-1}^i + S_t^{i-1} Pm_t^{i-1} - S_{t-dso\,(i-1)}^{i-1} Pm_{t-dso\,(i-1)}^{i-1} \tag{13.16}$$

$$R_t^i = R_{t-1}^i + S_t^i Pm_t^i - S_{t-dso\,(i)}^i Pm_{t-dso\,(i)}^i \tag{13.17}$$

$$ICR_t^i = P_t^i Pwip_t^i + Y_t^i (Pm_t^i - mr_t^i) \tag{13.18}$$

$$CumP_t^i = \sum_{k=1}^{k=t} S_k^i mr_k^i \tag{13.19}$$

13.3.6 Validation of the Behaviour Patterns of the Main Conwip SC Model Variables

In order to validate the behaviour patterns of the main variables of the CONWIP SC policy model, the following example similar to the beer game [23] was built, this time with variable demand. The Supply chain consists of four nodes (Figure 13.3) in series: a supplier, a first (1st) manufacturer, a second (2nd) manufacturer, and the distribution and retail channels. The behaviour pattern results are mainly presented for the global supply chain in order to study the impact of the centralised control and limited total inventory levels on performance measurers, while we treat the SC like a single cell. Customer demand is generated as a number from a normal distribution with mean assumed to be 4 units/week and standard deviation (*SD*) 4 units. Unit per container (UC^i) value is 1 and selected smoothing factor (α) is 0.2. The total number of production cards is 68. Initial conditions for WIP and FGI in firms are:

WIP: P_0^i = 8 beer cases, for i = 1 to n,

FGI: Y_0^i = 12 beer cases, for i = 1 to n.

The values adopted for the other operative parameters are included in Table 13.2, corresponding to Figure 13.3.

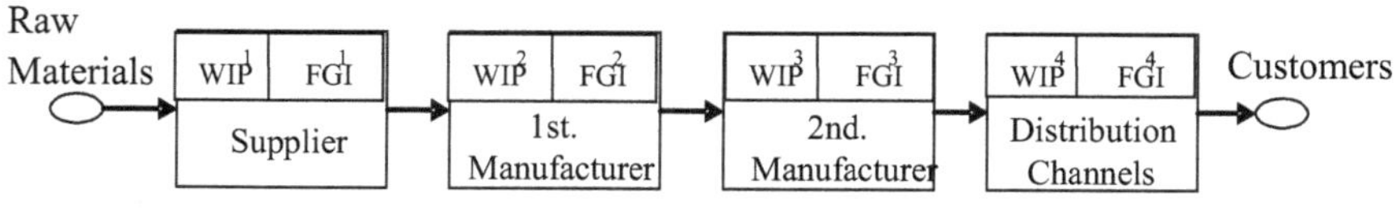

Figure 13.3. Sample SC selected for variable behaviour validation

Table 13.2. Model equations in fully-integrated SC *vs* CONWIP SC

	Supplier	1st Manufacturer	2nd Manufacturer	Distribution Channels
L^i *(weeks)*	2	2	2	2
MLP^i *(units)*	30	30	30	30

Figure 13.4 shows the behaviour of the orders placed by the factory and shipment to end-customers (throughput). In the scenario selected for this validation, the maximum load constraint does not limit the ordering policy; however the available production cards (Figure 13.5) could limit it (e.g. weeks 8 and 11). Either way, the ordering policy allows releasing enough part amounts to the supplier once they move to the FGI buffer of retailer on time for serving demand.

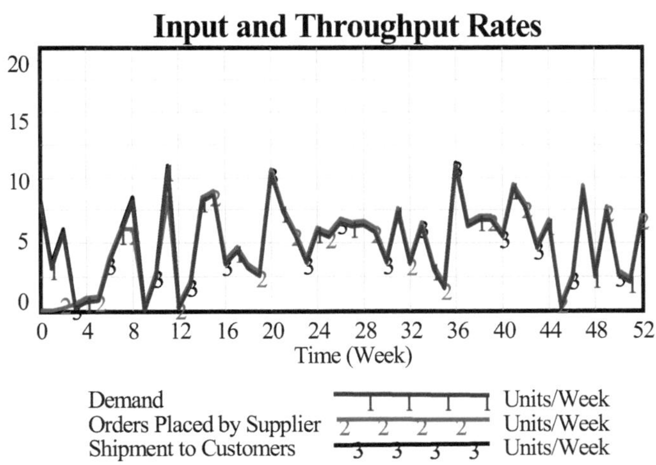

Figure 13.4. Input and throughput rates

Figures 13.4 and 13.5 also show that:

1. during the first three weeks it is not necessary to place orders due to established initial inventory amounts;
2. the condition *TNPC = APC+ total (WIP+FGI)* is completed; and
3. starting from the moment in which *WIP+FGI* in the chain reaches the steady state, as new orders arrive, sufficient number of cards were always available for releasing the necessary orders, in order to produce and meet all demand.

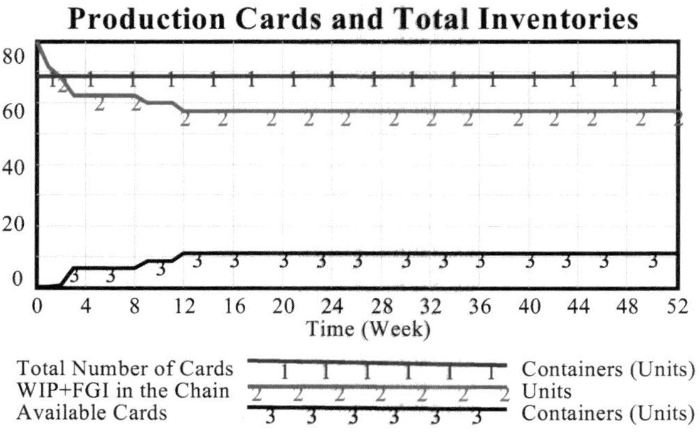

Figure 13.5. Production cards and total inventories

Figure 13.6 shows the behaviour of WIP+FGI per node. Inventories move between the firms in sequential form along the chain during the 52 weeks. This is a consequence of the existent push effect between firms of the SC (CONWIP systems characteristic) and the great parts load capacity settled in this example for all firms.

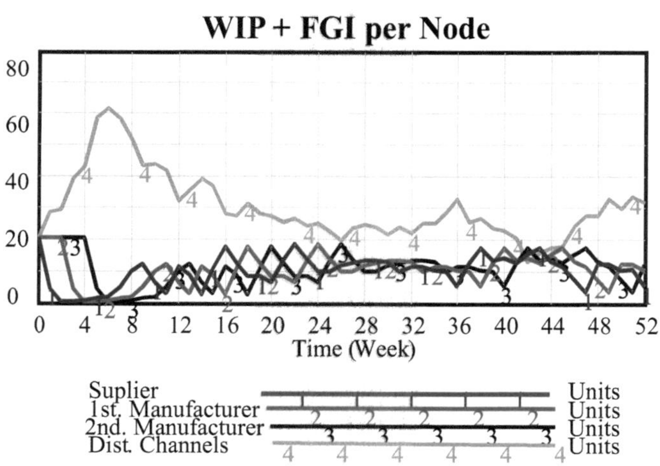

Figure 13.6. Total WIP + FGI per node

According to previous results obtained with this supply chain model, we have observed how it reproduces the main characteristics of a CONWIP control policy (according to [1]).

13.3.7 Simulation Study for the Comparison of SCM Policies

In this section we present a series of results comparing CONWIP SC with FI SC, and a set of performance metrics. These results are presented for both SCM policies and for the entire supply chain, consisting of four nodes (Factory, Distributor, Wholesaler and Retailer) as described by Sterman [21], in the article for the beer game. We remind the reader of the assumption made in this book: the retailer does not hold any backlog (due to the fact that end-customers will not wait in any scenario).

Keeping in mind that this study is carried out for a supply chain in which the final node serves end-customers, we set enough initial amount of WIP and FGI in each firm in order to ensure that the SC does not lose the initial sales, even for those scenarios where the production capacity constraint (MLP^i/L^i) may limit the ordering policies in some periods.

The simulation runs are for 52 weeks. In Table 13.3 we present the four demand situations considered in the study, and the initial values for WIP and FGI in all nodes.

Table 13.3. Specified parameters for final demand and initial inventory conditions

Demand (units/week) Normal Distribution Parameters					**Initial Conditions (Units)**	
Situation	Minimum	Maximum	Mean	Standard Deviation	WIP	FGI
1	0	8	4	4	8	12
2	0	8	4	3	8	12
3	0	8	4	2	4	8
4	0	8	4	1	3	7

Additionally, each of these situations is simulated varying the maximal load of parts to be processed in the firms (in all simulations we assume that we have the same value *MLP* for all firms), from 30 to 10 units. In summary, for each combination of demand SD and maximal load of parts (*MLP*), we will present the results of each SCM policy and for different performance metrics (see Figure 13.7).

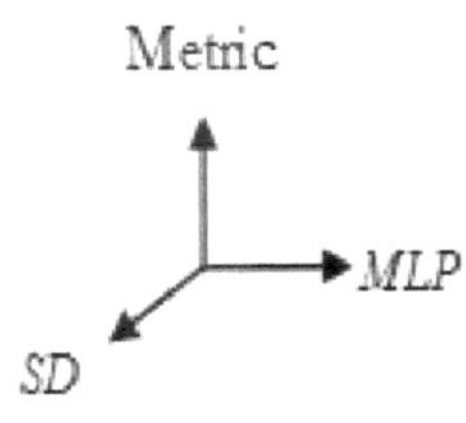

Figure 13.7. Dimensions of the simulation study carried out for each SCM Policy

A key issue in this study is that FI SC and CONWIP SC are compared under their optimal operating conditions for each particular scenario. These operating conditions are based on the achievement of two main goals settled in this Section: maximising throughput; and minimising total inventories in the chain under production capacity constraints (determined by the allowed maximal load of parts to be processed in a cycle time in firms).

To search the parameters values for each policy producing its best performance in each scenario, we have used a direct-search numerical optimisation technique which does not need to evaluate the gradient, and which is extremely suitable for the analysis of dynamics of complex nonlinear control systems. This technique is the modified Powell method [24,25], which is well known among direct-search methods, to obtain an extremely fast convergence.

Table 13.4 shows the equivalent variables to optimise and the parameters to search for each policy optimisation, and also shows the constraints of the system, equal for both cases. We have assigned weights to balance the numerical sizes of the variables to optimise, with the criteria of ensuring the same and high fill rates for both policies in the different scenarios. Later, the other operative and financial performance metrics will be compared for each case. In the next paragraph we will explain and justify the selection of the optimisation criteria.

Table 13.4. Optimisation criteria and CSM policy

	SCM Policy		
	Fully-Integrated (FI SC)	**CONWIP SC**	**Optimisation criteria**
Variables to optimise (payoff function)	Throughput	Throughput	Maximising
	Total Desired Production Orders	Available Production Cards	Minimising
	Total *WIP+FGI*	Total *WIP+FGI*	Minimising
Search parameters	SS , β_S , β_{SL}	*TNPC*	
Constraint	Max. Load (*MLP*)	Max. Load (*MLP*)	

Throughput is defined as shipment to end-customers. Desired production orders and WIP+FGI are the constituents of the base (target) stock in each node in case of FI SC policy. Similarly, available production cards and total WIP+FGI are the constituents of the total number of production cards in the chain for the CONWIP SC policy. In this way, for FI SC policy, we make sure that we minimise the base stock and, for CONWIP SC policy, we ensure that we minimise the total number of production cards, which are the main variables in the underlying management philosophy of each policy.

Table 13.5. Values for constant parameters in all simulations

Node parameters	Supplier	1st Manufact.	2nd Manufact.	Distrib. Channels	Units
- Operational:					
Cycle time, L^i	2	2	2	2	Week
Units per container, UC^i	1	1	1	1	Unit
Smoothing factor, α^i	0.2	0.2	0.2	0.2	Dmnl
- Financial:					
Price of a unit, Pm_t^i	120	208	303	383	\$/Unit
Weeks of sales Outstanding, $dso(i)$	2	2	2	2	Week
Profit margin, mr_t^i	0.5	0.4	0.3	0.2	Dmnl

Table 13.5 presents the values for parameters considered constant in all scenarios. The values obtained for search parameters are shown in Figure 13.8 and Table 13.6.

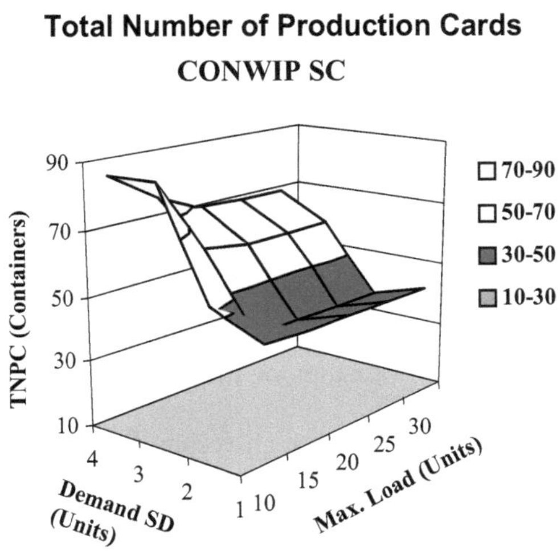

CONWIP SC	Demand SD			
Max. Load	**4**	**3**	**2**	**1**
30	68	60	39	44
25	68	60	39	44
20	69	60	39	44
15	75	62	39	44
10	85	85	52	45

Figure 13.8. Computed values for TNPC

Table 13.6. Computed values for SS, β_S and β_{SL}

FI SC Policy	Demand SD											
	4			3			2			1		
MLP	*SS*	b_S	b_{SL}	*SS*	b_S	b_{SL}	*SS*	b_S	b_{SL}	*SS*	b_S	b_{SL}
30	1.68	0.30	0.13	1.43	0.08	0.51	1.30	0.11	0.08	4.00	0.64	0.52
25	1.72	0.33	0.18	1.47	0.09	0.47	1.31	0.11	0.08	4.00	0.64	0.48
20	1.77	0.26	0.10	1.52	0.09	0.42	1.30	0.11	0.08	4.00	0.54	0.45
15	2.62	0.08	0.20	1.84	0.08	0.13	1.20	0.10	0.08	4.00	0.16	0.22
10	4.00	0.08	0.08	3.83	0.08	0.08	1.46	0.08	0.08	3.00	0.15	0.10

Regarding CONWIP SC policy, for common small values of the maximal load of parts to firms, the trend of the optimal *TNPC* is increasing. The reasons why the model computes high values for *TNPC* are:

1. The initial conditions value of *WIP*s and *FGI*s in firms impact on computing the total number of production cards.
2. In some periods, the ordered quantities may be shorter than the end-customer demand because the production capacity constraint of the factory delimits the ordering policy. When this happens, the chain will be able to serve demand only if there is enough accumulated inventory.

In consequence, in these scenarios the trend of the total *WIP+FGI* will be decreasing with time, and the trend of the available production card will be increasing. In the same sense, for the smaller values of demand SD, *TNPC* doesn't take the smallest values since the chain needs to accumulate enough inventories, especially when the available production cards delimits the ordering policy. In these cases, the maximal production capacity of the factory does not prevent holding high inventories (included *FGI*), avoiding lost sales.

For FI SC policy, note that the general trend in values obtained for SS is generally the same as for *TNPC*, with similar justification.

The following set of graphs shows the global comparative results for order, material and financial flow variables, and some measures of the chain financial performance.

Figure 13.9 shows how SCM policies obtain similar high results for service level (defined as fill rate, *i.e.* fraction of final customer incoming orders that is served by the SC) as was expected.

Results in Figure 13.10 show that, as allowed maximal load of parts to the factory decreases, the average orders placed by the factory also diminishes in both policies. Once again, this is due to the subsequent limits imposed by the maximal production capacity on the respective ordering policies. However, as demand SD decreases, the behaviour of the orders placed is analogous to the one of the total number of production cards and the safety stock.

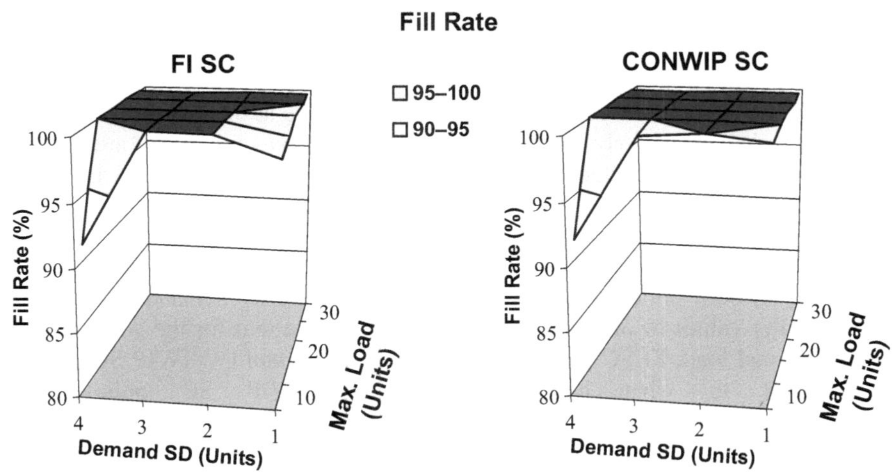

FI SC	Demand SD			
Max. Load	4	3	2	1
30	100	100	100	100
25	100	100	100	100
20	100	100	100	99.8
15	100	100	100	99.1
10	91.2	100	100	98.4

CONWIP SC	Demand SD			
Max. Load	4	3	2	1
30	100	100	100	100
25	100	100	100	100
20	100	100	100	100
15	100	100	100	100
10	91.5	99.7	100	99.5

Figure 13.9. Results for the service level

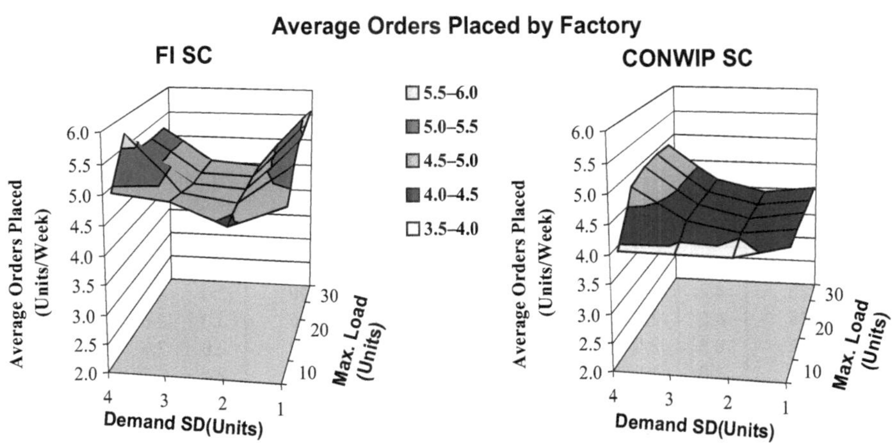

FI SC	Demand SD			
Max. Load	4	3	2	1
30	5.2	4.6	4.6	5.7
25	5.2	4.6	4.6	5.6
20	5.3	4.6	4.6	5.4
15	5.7	4.7	4.6	5.3
10	4.9	4.8	4.4	4.8

CONWIP SC	Demand SD			
Max. Load	4	3	2	1
30	4.9	4.2	4.0	4.2
25	4.9	4.2	4.0	4.2
20	4.9	4.2	4.0	4.2
15	4.8	4.2	4.0	4.2
10	3.9	3.9	3.9	4.1

Figure 13.10. Results for the average orders placed by the factory

On the other hand, results indicate that in all scenarios the CONWIP SC policy needs to place smaller average orders than the FI SC policy while they offer similar service levels, but this difference between them is bigger in scenarios with relative low maximal load and low demand SD. This result is due to the fact that CONWIP SC policy orders exclusively on the strength of the forecast while FI SC attempts to order to maintain a base stock (bigger than the forecast). This is a key issue for these results.

Figure 13.11 illustrates that, when the allowed maximal load of parts to the factory decreases, the standard deviation of orders placed by factory also diminishes in both policies. In the same sense, the smaller demand SD doesn't imply smaller values of orders placed variability. Notice that for the small values of the maximal load, FI SC presents smaller variability than CONWIP SC policy. In contrast to what happens in the CONWIP SC policy, the procurement/production rate constraints often delimit the quantities ordered by factory in the FI SC policy. This fact also controls variability of the orders placed. In general, CONWIP SC policy is less impacted overall by variability in demand.

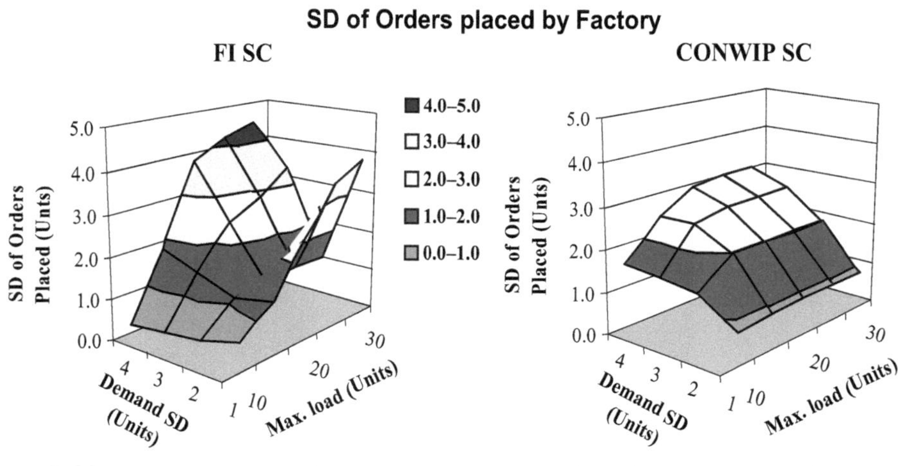

FI SC	Demand SD			
Max. Load	4	3	2	1
30	4.4	3.4	1.2	3.9
25	4.2	2.9	1.2	3.6
20	3.8	2.4	1.2	2.5
15	1.8	1.5	1.1	1.3
10	0.2	0.3	0.4	0.7

CONWIP SC	Demand SD			
Max. Load	4	3	2	1
30	3.0	2.6	2.0	0.9
25	3.0	2.6	2.0	0.9
20	2.9	2.6	2.0	0.9
15	2.4	2.5	2.0	0.9
10	1.5	1.5	1.4	0.8

Figure 13.11. Results for deviation standard of orders placed by factory

The possible bullwhip effect is substantially mitigated in CONWIP SC policy. The reasons for this are:

- centralised forecast and inventory management, avoiding a node to depend on another node;
- straight line of node's information to central control.

To summarise, both policies order whenever end-customers remove goods from the FGI of retailer, but parts release by CONWIP SC ordering policy is better (exactly) adjusted to the resulting mean (4.2 units in Figure 13.10) of demand when capacity and/or cards limitations appear.

The benefits achieved with the CONWIP SC policy can be clearly appreciated (Figure 13.12) in terms of less funds required to run that business model, especially during demand SD decreases, which represent substantial comparative financial savings. This is a consequence of the smallest inventory cash requirements needs, once the sales outstanding periods are all assumed to be equal (2 weeks).

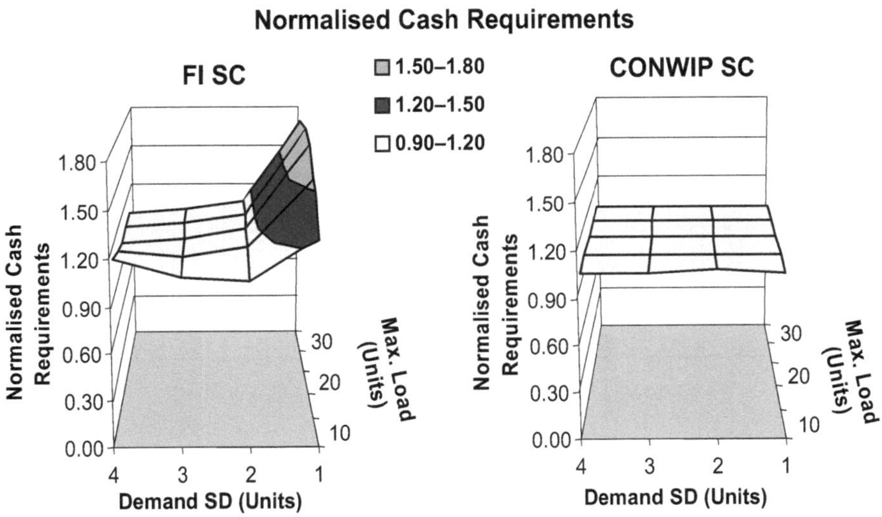

FI SC	Demand SD			
Max. Load	4	3	2	1
30	1.01	1.06	1.11	1.73
25	1.02	1.05	1.11	1.73
20	1.02	1.05	1.11	1.70
15	1.08	1.03	1.10	1.56
10	1.14	1.03	1.00	1.26

CONWIP SC	Demand SD			
Max. Load	4	3	2	1
30	1.00	1.00	1.00	1.00
25	1.00	1.00	1.00	1.00
20	1.00	1.00	1.00	1.00
15	1.00	1.00	1.00	1.00
10	1.00	1.00	1.02	1.00

Figure 13.12. Results for normalised cash requirements

Figure 13.13 shows one of the most significant differences between the two SCM policies. CONWIP SC policy reaches less average global inventory levels in all scenarios. This result shows an expected advantage of CONWIP SC policy over FI SC policy, that is, largest inventory efficiency giving similar service levels.

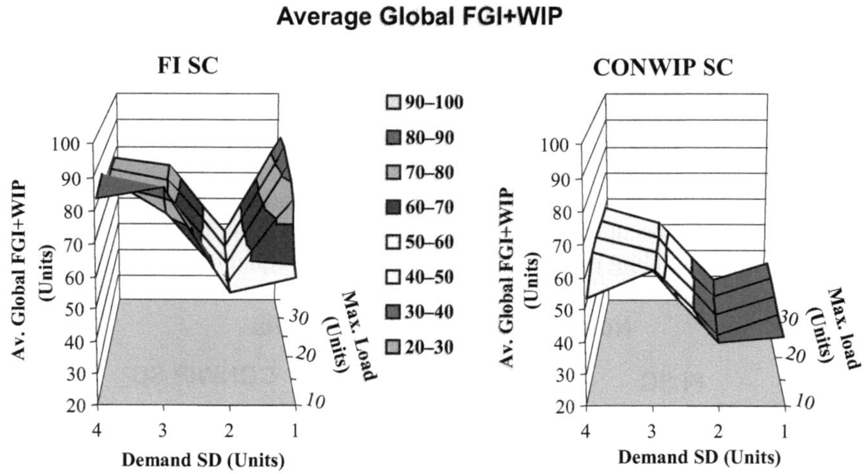

FI SC	Demand SD			
Max. Load	**4**	**3**	**2**	**1**
30	77	75	50	85
25	78	74	50	85
20	77	73	50	84
15	84	72	50	76
10	82	85	52	57

CONWIP SC	Demand SD			
Max. Load	**4**	**3**	**2**	**1**
30	59	53	31	38
25	59	53	32	38
20	59	53	31	38
15	60	53	31	38
10	51	59	36	38

Figure 13.13. Results for average global FGI+WIP

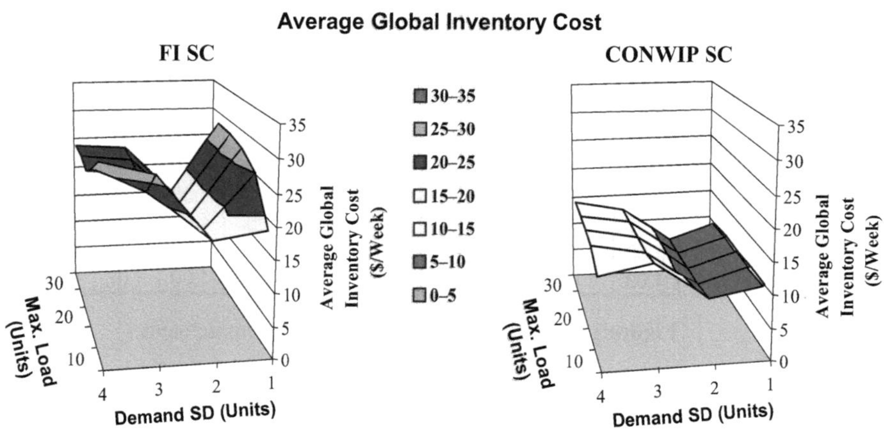

FI SC	Demand SD			
Max. Load	**4**	**3**	**2**	**1**
30	25	24	16	28
25	25	24	16	28
20	24	23	16	27
15	27	23	16	25
10	30	28	17	18

CONWIP SC	Demand SD			
Max. Load	**4**	**3**	**2**	**1**
30	15	13	8	10
25	15	13	8	10
20	15	13	8	10
15	15	13	8	10
10	13	14	9	10

Figure 13.14. Results for average inventory cost in the chain

Finally, with regard to cost of administration of inventories (holding more stock out penalties), CONWIP SC policy is more robust to incur in less costs in all situations (Figure 13.14). These results demonstrate more uniform materials flow, easier inventory control and better synchronisation between firms.

13.3.8 Conclusions of the Case Study for Comparison of SCM Policies

In this simulation study we evaluated the performance of the CONWIP control policy managing a supply chain in a variable environment, and compared that performance to the fully-integrated SC one. Based upon the results obtained, it appears that CONWIP SC policy represents a centralised control along the chain that may offer advantages in performance, compared to the decentralised FI SC. A CONWIP SC policy provides the following advantages over FI SC policy, when they offer similar service levels:

1. Easier control on WIP, since flow materials and parts are centrally controlled and limited.
2. CONWIP SC policy needs to place smaller average orders than FI SC especially in scenarios with relative low maximal load and low demand SD.
3. Generally the ordering policy in FI SC policy is more vulnerable as variability demand appear and load capacity in nodes is not very small. In CONWIP SC policy, centralised management of the demand and inventories mitigates the amplification of demand, allowing for the fact that the ordered quantities are adjusted to real flow material needs.
4. By CONWIP SC policy the supply chain obtain substantial comparative financial savings with regard to cash requirements, especially during demand SD decreases.
5. CONWIP SC policy offers potential shorter average FGI+WIP levels (larger efficiency) leading the supply chain to manage shorter average inventory costs.

In summary, although we may find some resistance to change in the firms acting as partners of the SC (see comments in Section 13.3), exploring the utilisation of CONWIP for SCM purposes could be a source of potential benefits in the near future.

13.4 References

[1] Spearman ML, Woodruff DL, Hopp WJ, (1990) CONWIP: A pull alternative to kanban. International Journal of Production Research, 28: 879–894.

[2] Hopp WJ, Spearman ML, (1996) Factory Physics. Chicago, USA: Irwin.

[3] Crespo Márquez A, Rubiano O. and Framinan JM, (2001) Benefits of the Internet for the Supply Chain Management. A Characterisation and simulation study. International Journal of Agile Manufacturing, 4(2):25–42.

[4] Spearman ML and Zazanis MA, (1992) Push and Pull Production systems: Issue and comparison. Operations Research, 40(3): 521–532.
[5] Muckstadt JA, Tayur SR, (1995a) A comparison of alternative kanban control mechanisms: I, Background and structural results. IIE Transactions, 27(1): 140–150.
[6] Muckstadt JA, Tayur SR, (1995b) A comparison of alternative kanban control mechanisms: II, Experimental results. IIE Transactions, 27(1): 151–161.
[7] Gstettner S, Kuhn H, (1996) Analysis of production systems Kanban and CONWIP. International Journal of Production Research, 34: 3253–3273.
[8] Tayur S, (1992) Properties of serial kanban systems. Queuing Systems, 12: 297–318.
[9] Tayur S, (1993) Structural properties and heuristic for kanban controlled serial lines. Management Science, 39: 1347–1368.
[10] Roderick LM, Phillips DT, Hogg GL, (1990) A comparison of order release strategies in production control systems. International Journal of Production Research, 30(3): 611–626.
[11] Bonvik AM, Gershwin SA, (1996) Beyond Kanban: Creating and analyzing lean shop floor control policies. Operations Research Center, MIT, Cambridge, MA 02139.
[12] Huang M, Wang D, Ip WH, (1998) Simulation study of CONWIP for a cold rolling plant. International Journal of Production Economics, 54(2): 257–266.
[13] Herer YT and Masin M, (1997) Mathematical programming formulation of CONWIP based production lines; and relationships to MRP. International Journal of Production Research, 35: 1067–1076.
[14] Wight OW, (1981) MRP II: Unlocking America' Productivity potential. Boston: CBI Publishing Co.
[15] Chase N, Aquilano N, (1985) Production and Operations Management. 4th Edition. Home-wood, Ill.: Irwin.
[16] Yang K, (2000) Managing a Flow Line with Single-kanban, Dual-Kanban or CONWIP. Production and Operations Management, 9(4): 349–366.
[17] Graves RJ, Konopka JM, Milne RJ, (1995) Literature review of material flow control mechanisms. Production planning and Control, 6(5): 395–403.
[18] Chen F, Ryan JK, Simchi-Levi D, (1999) The impact of exponential smoothing forecast on the bullwhip effect. Working paper, 1999, Northwestern University.
[19] Sanders NR, (1994) Forecasting practices en United-States corporations. Survey results. Interfaces, 24(2): 92–100.
[20] Tversky A, Kahneman D, (1974) Judgement under uncertainty. Heuristics and biases. Science, 185: 1124–1131.
[21] Sterman JD, (1989) Modelling managerial behaviour: Misperceptions of feedback in a dynamic decision making experiment. Management Science, 35(3): 321–339.
[22] Lineys JM, (1980) Corporate planning and policy design: A System Dynamics approach. MIT Press/Wright-Allen Series in System Dynamics.
[23] Sterman J, (1984) Instructions for Running the Beer distribution Game (D-3679). Sloan School of Management, MIT.
[24] Powell MJD, (1964) An Efficient method for finding the minimum of a function of several variables without calculating derivatives. Computer Journal, 7(2): 155–62.
[25] Powell MJD, (1968) On the calculation of orthogonal vectors. Computer Journal, 11(2): 302–304.
[26] Reddy R, Reddy S, (2001) Supply chains to virtual Integration. New York. McGraw-Hill.

[27] Rubiano Ovalle O, Crespo Marquez A, (2003) Exploring the utilisation of a CONWIP system for supply chain management. A comparison with fully integrated supply chains. International Journal of Production Economics, 83: 195–215.

14

Capacity Constraints Analysis for SCM

14.1 An Introduction to the Problem

In recent decades a lot of researchers have explored the relation between supply chain performance and capacity constraint. OR (Operational Research) methods have been used to support day-by-day decisions for batch sizing and job sequencing problems, while what-if analysis has been adopted as a decision support system in the field of supply chain reengineering as it enables the exploration of the impact of constrained capacity on the global performance of the whole structure.

Initial research in this specific topic can be found in Evans and Naim [1]. They investigated the relation between capacity constraint and supply chain performance in terms of demand amplification for a step increase in sales. They highlighted that in a traditional supply chain capacity constraints provide dampening of demand amplification, but in the meantime do not improve customer service levels with respect to the unconstrained case.

The aim of this chapter is to carry out a modelling exercise of this problem in such a manner as to contribute to this field of research, by answering the following research questions:

- How are different capacity levels related to the magnitude of demand amplification and customer service level?
- How are limited capacity impacts on supply chains characterised by different information sharing strategies?

This chapter is organised as follows; In Section 14.2 a review on recent supply chain modelling literature related to capacity constraints is presented. Section 14.3 reports the inventory control policy that is then modelled in the simulation study, the description of the adopted methodology and the mathematical formalism of the supply chain models. Performance metrics, experimental sets, data analysis and discussion are reported in Section 14.4. Finally, Section 14.5 provides conclusions and suggestions for future research.

14.2 Constrained Supply Chain Modelling in the Literature

In the field of supply chain modelling and analysis Riddals *et al.* [2] identify four main categories of methodologies, namely continuous time differential equation models, discrete time difference equation models, discrete event simulation systems and classical operational research (OR) methods. They state that OR techniques have their place at a local tactical level in the design of supply chains, while the implication of strategic design on supply chain performance is analysed by using the former methodologies.

In the constrained capacity supply chain analysis, optimisation approaches are typically used to solve batch sizing and job sequencing problems [3–6], while the impact of limited capacity on the global performance of the whole supply chain is generally assessed using control theory and what-if analysis, such as continuous simulation and discrete event simulation [1, 7–10].

Vlachos and Tagaras [3] present an optimisation procedure of a periodic review inventory system with regular and emergency replenishments. In their model the capacity of the emergency order quantity is constrained. Results show that the constraint on the capacity has a significant effect on the system performance under emergency ordering policies, especially when the review period and the regular replenishment lead time are long.

Bicheno *et al.* [4] present an algorithm to minimise inventory levels under constrained total capacity, determining optimal product-individual batch sizes and replenishment cycles under the constraint of limited available changeover time in an automotive steel supply network. The optimal batch sizes are found to be a function of all lead times, all demand volumes, all costs per unit, and the total available changeover time. The authors conclude that a change in the batch sizing policy combined with lead time reduction could provide savings in inventory of about 60% compared with the original situation.

Simchi-Levi and Zhao [5] develop a dynamic programming model to analyse a single product, periodic review, two-stage production-inventory system with a single capacitated manufacturer and a single retailer facing demand, under three different information strategies. The impact of information sharing on the manufacturer is studied as a function of the production capacity. They conclude that for the model with infinite production capacity the information sharing strategy has the same fill rate as the no information sharing strategy. When the production capacity is tightly constrained, the savings provided by information sharing with the optimal policy is relatively high.

Qi [6] develops a dynamic programming algorithm to study an integrated decision making model for a supply chain system where a manufacturer faces a price-sensitive demand and multiple capacitated suppliers. The goal is to maximise total profit by determining an optimal selling price and at the same time acquiring enough supplying capacity. He concludes that, when the sourcing information is not available, it is better to make a conservative production plan, and when the market demand information is unknown, it is better to make an aggressive production plan.

Evans and Naim [1] develop a continuous time differential equation model of a three-tier Forrester [11] supply chain. Eight combinations of three capacity levels are used to study the response of constrained traditional serially-linked structures for a step input in customer demand. Their most notable result is that the unconstrained case is not the highest ranked system in the simulations performed, as it is characterised by the presence of demand amplification, commonly know as bullwhip effect [12–19]. On the contrary, they highlight that constrained supply chains may produce improved system performance at echelon level in terms of demand amplification, but this does not mean improved customer levels.

Gavirneni *et al.* [7] perform an infinitesimal perturbation analysis to address a periodic review inventory control problem in three two-echelon supply chains, in order to explore the trade-offs between inventories, capacities, and information. One of their conclusions is that information is more beneficial if the supplier's capacity is high as compared to when it is low. Helo [8] discusses, by way of using a system dynamic simulation, the trade-off between capacity utilisation and lead times. The analysis recommends smaller order sizes, echelon synchronisation and capacity analysis as methods of improving the responsiveness of a supply chain.

Disney and Grubbström [9] use z-transforms and probability density functions to analyse the economical impact of order and inventory-related cash flows resulting from a generalised order-up-to policy, considering costs associated with the production order rate within and above a capacity constraint. They conclude that the classical order-up-to policy is no longer optimal when a broader range of costs is considered in the objective function. It is shown that incorporating proportional controllers in the two feedback loops is economically desirable for a particular scenario and a particular set of cost functions

Wikner *et al.* [10] study, via a System Dynamics tool, the properties of the make-to-order environment with finite capacity under the Automatic Pipeline Inventory and Order Based Production Control System archetype, commonly known as APIOBPCS replenishment rule [20]. It is shown that, under capacity constraints, existing production planning and control systems should accommodate a comparator to utilise the difference in the target and actual backorders in the ordering rule. The authors highlight the fact that by developing an ordering policy that accounts for capacity flexibility, plus the feedback monitoring of the backlog state, it is possible to ensure lead time expectations.

In this chapter a dynamic simulation model is presented. The objective of this model is to investigate the performance of multi-tier supply chain structures, analysed under different information sharing strategies and capacity constraint conditions.

14.3 Modelling the Constrained Supply Chain

14.3.1 Inventory Control Policy Models

There are several classifications for the policies that regulate the flow of materials within the supply chain. Swaminathan *et al.* [21] distinguish two main types of inventory control: the *centralised* and the *decentralised.* The centralised policy takes into account the inventory levels in the supply chain as a whole. The ability to access information on inventory levels at other trading partners in the supply chain is a fundamental requirement for implementing the centralised management. A typical centralised supply chain is the Vendor Managed Inventory [22–26]. In the decentralised inventory control the replenishment rule takes into account only the information on local inventory status.

A further classification of inventory control policies is based on the replenishment rule: the periodic review/order-up-to/base-stock policy, and the continuous review/reorder point/order quantity model [27]. In a base-stock policy the review period is fixed, and the size of the order is such that the inventory position is raised up to a target level. A modification to the classical order-up-to policy is obtained by introducing a proportional controller in the inventory position feedback loop [28]. This kind of inventory control policy is called smoothing replenishment rule.

In this work the capacity constraint condition is investigated in two decentralised and one centralised four-tier serially-linked supply chain, under periodic review/smoothing order-up-to/base-stock policy.

14.3.2 Model Notation

The supply chains modelled in this work are single product models with no bill of material between echelons. In Figure 14.1 the item flow is represented.

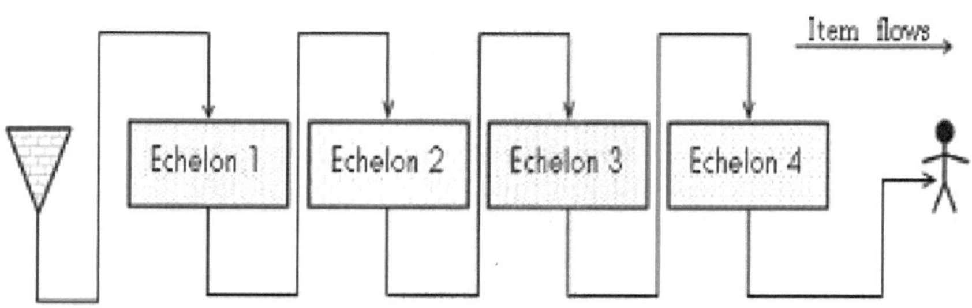

Figure 14.1. Supply chain item flow

The nomenclature of the model variables is now presented, where material flow variables are separated from information flow variables and model parameters, as follows:

Material variables:

Wip_t^i	WIP (includes incoming transit units) at echelon *i* at time *t*
Inv_t^i	Inventory of finished materials at echelon *i* at time *t*
S_t^i	Units/orders finally shipped from echelon *i* at time *t*
F_t^i	Throughput at echelon *i* at time *t*

Information variables:

$\hat{d}_t^i$	Demand forecast at echelon *i* at time *t*
$\hat{d}_t^{cust}$	Customer demand forecast at time *t*
d_t^{cust}	Customer demand at time *t*
R_t^i	Replenishment order quantity at echelon *i* at time *t* (amount of items that would be required with no capacity constraint)
Rcc_t^i	Actual replenishment order quantity at echelon *i* at time *t* (amount of items required under capacity constraint condition)
B_{t-1}^i	Existing backlog of orders at echelon *i* at time *t*
$TInv_t^i$	Target inventory at echelon *i* at time *t*
$TWip_t^i$	Target work in process at echelon *i* at time *t*
$VirtInv_t^i$	Virtual inventory at echelon *i* at time *t*
$VirtWip_t^i$	Virtual work in process at echelon *i* at time *t*
$TVirtInv_t^i$	Target virtual inventory at echelon *i* at time *t*
$TVirtWip_t^i$	Target virtual work in process at echelon *i* at time *t*
$\sigma^2_{Rcc^i}$	Order quantity variance at echelon *i*
μ_{Rcc^i}	Order quantity mean value at echelon *i*
$\sigma^2_{d^{cust}}$	Customer demand variance
$\mu_{d^{cust}}$	Customer demand mean value

Parameters:

α Forecast smoothing factor

Δ Final steady-state demand

cf Capacity factor

T_p^i Physical production/distribution lead time at echelon i (incoming transit time from supplier plus the production lead time)

T_c^i Cover time for the inventory control at echelon i

T_y^i Smoothing inventory parameter at echelon i

T_w^i Smoothing work in process parameter at echelon i

14.3.3 The Decentralised Model

The decentralised base-stock supply chain (model A) presented in this work is a serially-linked four-echelon supply chain, in which trading partners use a smoothing replenishment rule. Each echelon only receives information on local stock, local work in process levels, and local sales. The retailer forecasts customer demand on the basis of market time series, and the remaining echelons only take into account for their replenishment downstream incoming orders.

Equations 14.1, 14.2 and 14.3 represent the state variables of the model:

$$Wip_t^i = Wip_{t-1}^i + S_t^{i-1} - F_t^i \tag{14.1}$$

$$Inv_t^i = Inv_{t-1}^i + F_t^i - S_t^i \tag{14.2}$$

Work in process (Equation 14.1) and Inventory (Equation 14.2) describe the physical flow of items in downstream direction. Note that at every echelon the shipments sent by the supplier S_t^{i-1} immediately become work in process.

$$B_t^i = B_{t-1}^i + Rcc_t^{i+1} - S_t^i \tag{14.3}$$

Backlog (Equation 14.3) is representative of service level for each tier. Backlogging is allowed as a consequence of stockholding; in each echelon the backlog will be fulfilled as soon as on-hand inventory becomes available.

$$S_t^i = \min(Rcc_t^{i+1} + B_{t-1}^i; Inv_{t-1}^i + F_t^i) \tag{14.4}$$

Equation 14.4 expresses the dynamic of delivered orders.

$$F_t^i = S_{t-T_p}^{i-1} \tag{14.5}$$

Equation 14.5 models the production/delivery lead time delay, represented by the parameter *Tp*.

$$\hat{d}_t^i = \alpha Rcc_{t-1}^{i+1} + (1-\alpha)\hat{d}_{t-1}^i ; 0 < \alpha \le 1, \forall i \ne 4 \tag{14.6}$$

$$\hat{d}_t^{cust} = \alpha d_{t-1}^{cust} + (1-\alpha)\hat{d}_{t-1}^{cust} ; \ 0 < \alpha \le 1 \tag{14.7}$$

$$\hat{d}_t^i = \hat{d}_t^{cust} \quad \forall i = 4 \tag{14.8}$$

Equations 14.6 and 14.7 represent the exponential smoothing formula to forecast demand [29]. The forecast smoothing factor α represents the weighting factor of the exponential smoothing rule. The value of α is between 0 and 1. The higher the value of α, the greater is the weight placed on the more recent demand levels. The lower the α value the greater is the weight given to demand history in forecasting future demand.

Two equations are adopted to take into account the forecast on customer orders $\hat{d}_t^{cust}$ and the forecast on orders placed by tiers $\hat{d}_t^i$. Equation 14.8 shows that only the tier next to the final customer carries out a forecast based on market demand, while at the upstream stages the input demand data is given by Equation 14.6.

The adopted periodic review/smoothing order-up-to/base-stock replenishment rule (Equation 14.9) is the APIOBPCS. This order policy is expressed as "let the production targets be equal to the sum of a forecast of perceived demand, plus a fraction (1/Ty) of the inventory discrepancy between actual and target levels of finished goods, plus a fraction (1/Tw) of the discrepancy between target WIP and actual WIP" [14].

$$R_t^i = \hat{d}_t^i + \frac{1}{T_w^i}(TWip_t^i - Wip_t^i) + \frac{1}{T_y^i}(TInv_t^i - Inv_t^i) \tag{14.9}$$

Equation 14.10 models the non-negativity condition for the replenishment order quantity:

$$R_t^i \ge 0, \forall i \tag{14.10}$$

There are several ways for implementing a capacity constraint condition within a supply chain model; a limitation could be placed on the order rate or in the order acceptance channel. Evans and Naim [1] suggest that constraining the order quantity appears to be the realistic placement within the supply chain. Equation 14.11 expresses the actual replenishment order quantity, which is the amount of items required under capacity constraint condition:

$$Rcc_t^i = \min(R_t^i; cf) \qquad (14.11)$$

The order quantity of every tier cannot exceed the capacity factor *cf*, which is computed as a multiple of δ, the actual marketplace demand after the step-shaped increase.

Target inventory (in Equation 14.12) is updated every period according to the covering time and the new demand forecast:

$$TInv_t^i = \hat{d}_t^{\,i} \, T_c \qquad (14.12)$$

Target orders are placed in the pipeline on the basis of the demand forecast and production/delivery lead time (Equation 14.13):

$$TWip_t^i = \hat{d}_t^{\,i} \, T_p \qquad (14.13)$$

14.3.4 POS Decentralised Model

In the POS decentralised base-stock supply chain (model B) all echelons base their inventory policy on local stock, local work in process levels, local sales, downstream incoming orders and the actual marketplace demand. This structure is modelled through Equations 14.1–14.8, 14.10–14.13 and 14.4. The order decision rule implemented in the POS decentralised base-stock (Equation1 4.4) takes into account the conjoint use of the market demand forecast (Equation 14.7), based on the end consumer order rate, and the demand forecast at echelon *i* (Equation 14.6), based on the orders placed by the subsequent stage. The customer demand forecast is directly included into the replenishment rule, while the forecast on the order incoming by echelon *i*+1 is used to compute Target Work in Process (Equation 14.13) and Target Inventory (Equation 14.12), as in the decentralised model:

$$R_t^i = \hat{d}_t^{\,cust} + \frac{1}{T_w^i}(TWip_t^i - Wip_t^i) + \frac{1}{T_y^i}(TInv_t^i - Inv_t^i) \qquad (14.14)$$

14.3.5 Centralised Model

The inventory policy for the centralised base-stock supply chain (model C) is based on local stock, local work in process levels, local sales, downstream incoming orders, actual marketplace demand, inventory information and work in process data incoming from the downstream trading partner. The model is described by Equations 14.1–14.5, 14.7, 14.10, 14.11, 14.15–14.19. Equation E.15 represents the periodic review order quantity for the centralised base-stock supply chain:

$$R_t^i = \hat{d}_t^{cust} + \frac{1}{T_w^i}(TVirtWip_t^i - VirtWip_t^i) + \frac{1}{T_y^i}(TVirtInv_t^i - VirtInv_t^i) \tag{14.15}$$

The variable Virtual Inventory for an individual echelon i is the sum of the local Inventory plus all Inventories of subsequent echelons (Equation 14.16):

$$VirtInv_t^i = \sum_{j=i}^{4} Inv_t^j \tag{14.16}$$

The variable Virtual Work In Process for echelon i is given by orders-in-the-pipeline at stage i plus the sum of work in process of all downstream echelons (Equation 14.17):

$$VirtWip_t^i = \sum_{j=i}^{4} Wip_t^j \tag{14.17}$$

Target Virtual Inventory (Equation 14.18) in echelon i depends on the forecasted marketplace demand and on the sum of the local and subsequent tiers' cover times for the inventory control:

$$TVirtInv_t^i = \hat{d}_t^{cust} \sum_{j=i}^{4} T_c^j \tag{14.18}$$

Target Virtual Work In Process (Equation 14.19) in echelon i depends on the forecasted marketplace demand and on the sum of the local and subsequent stages' physical production distribution lead time:

$$TVirtWip_t^i = \hat{d}_t^{cust} \sum_{j=i}^{4} T_p^j \tag{14.19}$$

14.4 Performance Metrics, Experiments and Discussion

14.4.1 Supply Chain Performance Metrics

In this study the supply chain performance is evaluated through the following metrics: Order Rate Variance Ratio (ORVrRatio), also known as Bullwhip Magnitude, Average Inventory, and Backlog.

The Order Rate Variance Ratio (Equation 14.20) was proposed by Chen *et al.* [13] to quantify the bullwhip effect.

$$ORVrRatio^i = \frac{\sigma^2_{Rcc^i} / \mu_{Rcc^i}}{\sigma^2_{d^{cust}} / \mu_{d^{cust}}} \tag{14.20}$$

Equation 14.20 is a smart and concise quantification of the order rate instability. One value of Order Rate Variance Ratio is computed for each echelon in the chain. A geometric or exponential increase of ORVrRatio in upstream direction in the supply chain is representative of the transmission of bullwhip effect [30]. The values of Order Rate Variance Ratio are interpolated along each chain and the curve slopes are calculated. The *slope* of Order Rate Variance Ratio is a single value, which is indicative of the extent of bullwhip effect propagation and inventory instability along a given supply chain structure.

A complementary measure of supply chain performance is the *Average Inventory* (Equation 14.21), computed as the mean value of Inventory at echelon i over the simulation time span T:

$$Average\ Inventory^i = \frac{1}{T}\sum_{t=0}^{T} Inv^i_t \tag{14.21}$$

A concise measure of multi-echelons system performance related to stock levels is the *Global Average Inventory* (GAI), computed as the sum of Average Inventory values over the four tiers.

As a customer service level measure, Backlog (Equation 14.3) is adopted to quantify the accumulation of unfilled orders. The Backlog is evaluated every single step and the time series reproduces the supply chain customer service level history. To associate a customer service level indicator to each supply chain and concisely compare different scenarios, an additional measure is used: the *Average Backlog* (Equation 14.22):

$$Average\ Backlog^i = \frac{1}{T}\sum_{t=0}^{T} B^i_t \tag{14.22}$$

14.4.2 Experimental Sets: Assumption and Parameter Vectors

For each base-stock model six capacity constraint levels are studied. *The capacity factor cf is modelled as a function of the final marketplace steady state demand δ.* This relation between capacity constraints and customer demand is also assumed in Evans and Naim [1] and Simchi-Levi and Zhao [5]. In this work the six selected levels of capacity factor are [δ; 1.5δ; 2δ; 3δ; 4δ; ∞].

The experimental sets are characterised by:

- The simulation runs are for a total of 52 time units, with constant time interval equal to Δt=0.25 time units;
- Marketplace demand is assumed to be 4 units per time unit, until there is a pulse at t=5, increasing the demand value up to 8 units per time unit;
- The values of the parameter vector [α; cf; T_p; T_c; T_y; T_w] elements are: forecast smoothing factor α=0.5; physical production/distribution lead time T_p=2; cover time for the inventory control T_c=3; smoothing inventory parameter T_y=3; smoothing work in process parameter T_w=3. The smoothing inventory parameter T_y and the smoothing work in process parameter T_w are chosen on the basis of the empirical formula $T_y=T_w=1+T_p$, [31]. Note that the Deziel and Eilon [32] smoothing parameter configuration is used.
- The state value vector at t=0 [Wip_0^i; Inv_0^i; B_0^i] is as Sterman's configuration [33].
- For echelon 1 (manufacturer) no replenishment lead time is considered.

The following sub-section will report data analysis for the experimental sets.

14.4.3 Data Analysis

For the decentralised base-stock supply chain (model A), the POS decentralised base-stock supply chain (model B), and the centralised base-stock supply chain (model C) all the performance measures are grouped for capacity factor *cf*.

The values of Order Rate Variance Ratio, measured in individual echelons, are reported in Table 14.1. The slopes of the interpolations are shown in italic in the last row of each data set. Echelon 1 is excluded from the slope computation, due to the fact that it structurally differs from the others as its replenishment lead time is null.

Table 14.1. Order rate variance ratio

Order rate variance ratio – slope						
	cf δ			*cf 1.5δ*		
	A	*B*	*C*	*A*	*B*	*C*
4	1.05	1.05	1.05	6.77	6.33	5.11
3	1.15	1.12	1.02	11.82	9.23	5.24
2	1.24	1.15	1.01	13.59	10.28	4.64
1	0.73	0.56	0.48	6.99	5.40	1.81
slope	*0.09*	*0.04*	*-0.02*	*3.40*	*1.97*	*-0.23*
	cf 2δ			*cf 3δ*		
	A	*B*	*C*	*A*	*B*	*C*
4	9.68	8.82	5.47	9.32	8.69	4.93
3	19.61	16.04	8.80	41.20	27.02	11.20
2	26.03	18.68	7.07	52.04	37.18	12.49
1	13.02	10.14	2.99	27.17	16.93	3.79
slope	*8.17*	*4.92*	*0.79*	*21.36*	*14.24*	*3.78*
	cf 4δ			*cf ∞*		
	A	*B*	*C*	*A*	*B*	*C*
4	9.35	8.71	4.93	9.51	8.71	4.93
3	48.75	27.03	11.20	47.99	27.03	11.20
2	85.19	40.44	12.49	129.1	40.44	12.49
1	46.89	17.87	3.79	82.99	17.87	3.79
slope	*37.92*	*15.86*	*3.78*	*59.80*	*15.86*	*3.78*

For each echelon the values of Average Inventory are reported in Table 14.2. The Global Average Inventory is shown in italic.

Table 14.2. Average inventory

Average inventory							
	Cf δ				***Cf 1.5δ***		
	A	*B*	*C*		*A*	*B*	*C*
4	7	7	8		18	18	18
3	8	8	8		12	15	16
2	8	8	8		13	14	16
1	8	8	8		14	14	16
GAI	*31*	*31*	*32*		*57*	*61*	*66*
	cf 2δ				***cf 3δ***		
	A	*B*	*C*		*A*	*B*	*C*
4	19	19	19		19	19	19
3	18	18	18		29	22	19
2	18	18	17		30	22	18
1	20	20	17		31	22	17
GAI	*75*	*75*	*71*		*109*	*85*	*73*
	cf 4δ				***cf ∞***		
	A	*B*	*C*		*A*	*B*	*C*
4	19	19	19		19	19	19
3	33	22	19		32	22	19
2	49	23	18		68	22	18
1	43	22	17		77	22	17
GAI	*144*	*86*	*73*		*196*	*85*	*73*

The values of backlog measured in individual echelons are plotted over the entire time horizon in Figure 14.2.

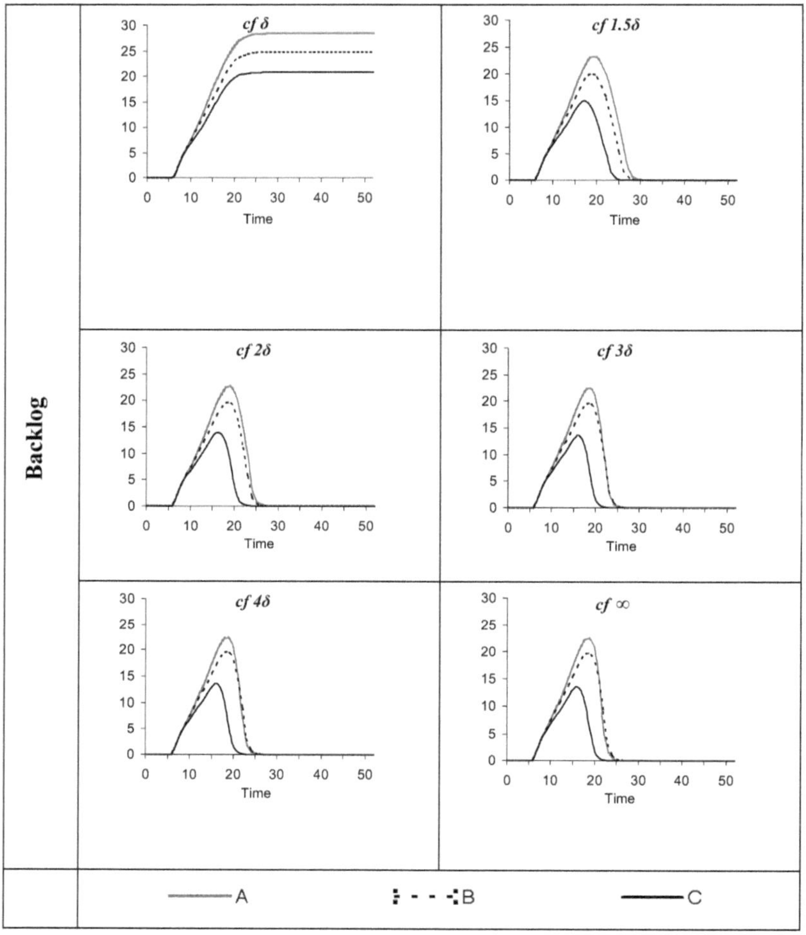

Figure 14.2. Backlog

The average backlog measures for model A, B, and C are shown in Table 14.3.

Table 14.3. Average backlog

	Average backlog					
	cf δ	*cf 1.5δ*	*cf 2δ*	*cf 3δ*	*cf 4δ*	*cf ∞*
A	20.86	5.45	4.54	4.16	4.06	4.03
B	18.33	4.49	3.92	3.79	3.79	3.79
C	15.58	2.95	2.36	2.17	2.17	2.17

14.4.4 Discussion

The first research question addressed by this work was to explore how different capacity levels are related to the magnitude of demand amplification and customer service level.

Using actual inventory levels as customer service level, Evans and Naim [1] provide a qualitative insight into the relationship between customer care and capacity constraint. A limitation of this metric is that it does not provide a direct link between levels of capacity constraint and extent of customer service level. In this work this limitation is overcome through the use of backlog. Monitoring this variable contributes to answer the first research question. When the capacity factor is equal to the final steady demand, a step increase in demand causes a diminution of service level, and capacity saturation impedes to recover the accumulation of previous unfulfilled orders.

Backlog values do not significantly improve for increasing values of *cf*, including the unconstrained capacity case. This result suggests that in a decentralised supply chain an increment of production capacity does not necessarily cause an improvement in customer service.

In the aforementioned study, Evans and Naim [1] state that the traditional supply chain under unconstrained capacity is not the highest ranked system in terms of demand amplification dampening.

In the first supply chain analysed in this work (model A) the slope values of Order Rate Variance Ratio increase monotonously as the capacity factor steps up. The same performance trend is seen in the Global Average Inventory. These results confirm Evans and Naim's conclusions [1] for a decentralised structure: the capacity constraint provides a general improvement of process performance within the multi-echelon system for a step input in demand, in terms of demand amplification and supply chain stability.

The relation between supply chain performance improvement and capacity limitations is due to the fact that the capacity constraint limits the overestimated forecast on the real step-shaped marketplace demand along the chain, as it dampens the order quantities.

Note that the bullwhip reduction associated with the constrained capacity is an apparent improvement of the performance of the supply chain.

In this study, it is assumed that one product is processed with no bill of material, that is a one to one ratio of raw material, and in particular the partnership between tiers is based on structured and exclusive contracts. However, in the real business world, when production capacity is saturated (fully utilised) a trading partner would search for additional capacity by enlarging its portfolio with unstructured contracts. This is a double risk, as it can lead to satisfying at a higher cost a demand that is amplified by information distortion. The traditional supply chain is more inclined to extreme demand amplification along the whole supply chain [9], and a direct consequence of the bullwhip effect is a remarkable increment in tiers' order rates that correspond to a greater production capacity need, and it increases the chance of incurring the aforementioned double risk.

In the presence of demand amplification, the capacity constraint can lead to satisfying at a higher cost, particularly in a supply chain with unstructured contracts, an over-estimated market demand.

The second research question was to understand how limited capacity impacts on supply chains characterised by different information sharing strategies.

A general result is that the POS decentralised (model B) and the centralised supply chain (model C) outperform the decentralised structure, both in terms of process metrics and customer service.

As in model A, for models B and C the slope of Order Rate Variance Ratio and the value of Global Average Inventory increase as *cf* grows, but the magnitude of this increment is not significant. This is particularly evident in the decentralised supply chain.

The "side effect" of demand amplification dampening, provided by the capacity constrained condition, has a lower impact on supply chains characterised by information sharing strategies. This effect is due to the fact that the redesign of information patterns is one of the most effective solving approaches to demand amplification phenomenon and supply chain instability. Information sharing practices overshadow the capacity limitation smoothing effect on demand amplification.

Analysing the case of capacity factory equal to the final steady-state demand, it is shown that capacity saturation impedes to recover the accumulation of backlogged unfulfilled orders, also in the information sharing supply chains. Regardless of the persistence of unfulfilled orders, the average backlog decreases as the level of information sharing increases. This reduction trend is seen in all the experimental sets.

The results suggest that the effect of limited capacity on information sharing supply chains is significantly reduced with respect to classical decentralised base-stock multi-echelons in terms of demand amplification dampening and customer service level.

Under collaborative practices, bullwhip dampening and inventory stability provided by information sharing increase the ability of the structure to avoid the risk of satisfying at a higher cost a demand that is amplified by information distortion.

In the supply chain re-engineering field, expanding the production capacity is a local approach commonly used in industry to manage increasing incoming orders. When the increased demand is due to the presence of information distortion along the chain, this solution converts into an amplifier of bullwhip effect.

In the strategic capacity management, one priority is to eliminate the information distortion in the supply chain, in order to dimension the production/distribution channel capacity with relation to the actual marketplace demand.

14.5 Concluding Remarks

The aim of this chapter was to explore the relationship between constrained capacity and supply chain performance. The effect of six capacity constraint levels was studied in two decentralised and one centralised four-tier serially-linked supply chain, under periodic review/smoothing order-up-to/base-stock policy.

The supply chain performance metrics used in this work were Order Rate Variance Ratio, Average Inventory, and Backlog.

The main results of this study are:

- In a decentralised supply chain an increment of production capacity does not necessarily cause an improvement in customer service.
- In the presence of demand amplification, the capacity constraint can lead to a double risk, satisfying at a higher cost an over-estimated market demand. A supply chain with unstructured contracts is more affected by this negative phenomenon.
- Under collaborative practices, bullwhip dampening and inventory stability provided by information sharing increase the ability of the structure to avoid the double risk.
- In the strategic capacity management, one priority is to eliminate the information distortion in the supply chain, in order to dimension the production/distribution channel capacity with relation to the actual marketplace demand.

Future research will involve studies on the limited capacity production/distribution under several real supply chain conditions, through the use of complementary methodological approaches, such as discrete event simulation.

14.6 References

[1] Evans GN, Naim MM, (1994) The dynamics of capacity constrained supply chains. Proceedings of International System Dynamics Conference, Stirling, Scotland, pp. 28–35.

[2] Riddals CE, Bennett S, Tipi NS, (2000) Modelling the dynamics of supply chains, International Journal of Systems Science, 31(8): 969–976.

[3] Vlachos D, Tagaras G, (2000) An inventory system with two supply modes and capacity constraints, International Journal of Production Economics, 72(1): 41–58.

[4] Bicheno J, Holweg M, Niessman J, (2001) Constraint batch sizing in a lean environment. International Journal of Production Economics, 73(1): 41–49.

[5] Simchi-Levi D, Zhao Y, (2003) The value of information sharing in a two-stage supply chain with production capacity constraints. Naval Research Logistics, 50(8): 888–916.

[6] Qi X, (2007) Order splitting with multiple capacitated suppliers. European Journal of Operational Research, 178(2): 421–432.

[7] Gavirneni S, Kapucinski R, Tayur S, (1999) Value of information in capacitated supply chains. Management Science, Vol. 45(1): 16–24.

[8] Helo PT, (2000) Dynamic modelling of surge effect and capacity limitation in supply chains. International Journal of Production Research, 38(17): 4521–4533.

[9] Disney SM, Grubbström RW, (2004) Economic consequences of a production and inventory control policy. International Journal of Production Research, 42(17): 3419–3431.
[10] Wikner J, Naim MM, Rudberg M, (2007) Exploiting the Order Book for Mass Customised Manufacturing Control Systems with Capacity Limitation. IEEE Transactions on Engineering Management, 54(1): 145–155.
[11] Forrester JW, (1961) Industrial Dynamics. Sloan School of Management. Cambridge: MIT Press.
[12] Lee HL, Padmanabhan V, Whang S, (1997) Information distortion in a Supply Chain: The Bullwhip effect. Management Science, 43(4): 546–558.
[13] Chen F, Drezner Z, Ryan JK, Simchi-Levi D, (2000) Quantifying the bullwhip effect in a simple Supply Chain: the impact of forecasting, lead-times and information. Management Science, 46(3): 436–43.
[14] Disney SM, Towill DR, (2003a) The effect of vendor managed inventory (VMI) dynamics on the Bullwhip Effect in Supply Chains. International Journal of Production Economics, 85(2): 199–215.
[15] Disney SM, Towill DR, (2003b) On the bullwhip and inventory variance produced by an ordering policy. The International Journal of Management Science, 31(3): 157–167.
[16] Chatfield DC, Kim JG, Harrison TP, Hayya JC, (2004) The bullwhip effect – Impact of stochastic lead time, information quality, and information sharing: A simulation Study. Production and Operations Management, 13(4): 340–353.
[17] Holweg M, Disney SM, (2005) The evolving frontiers of the bullwhip problem. Proceedings of the conference EurOMA: Operations and global competitiveness, Budapest Hungary June 19–22, pp. 777–716.
[18] Geary S, Disney SM, Towill DR (2006) On bullwhip in supply chains - historical review, present practice and expected future impact. International Journal of Production Economics, 101(1): 2–18.
[19] Miragliotta G, (2006) Layers and mechanisms: A new taxonomy for the bullwhip effect. International Journal of Production Economics, 104(2): 365–381.
[20] John S, Naim MM, Towill DR, (1994) Dynamic analysis of a WIP compensated decision support system. International Journal of Management Systems Design, 1(4): 283–297.
[21] Swaminathan JM, Smith SF, Sadeh NM, (1998) Modelling Supply Chain Dynamics: A Multiagent Approach. Decision Sciences, 29(3): 607–632.
[22] Waller M, Johnson ME, Davis T, (1999) Vendor-managed inventory in the retail supply chain. Journal of Business Logistics, 20(1): 183–203.
[23] Disney SM, Towill DR (2002) A procedure for the optimisation of the dynamic response for a Vendor Managed Inventory supply chain. Computers and Industrial Engineering: An International Journal, 43(1–2): 27–58.
[24] Kuk G, (2004) Effectiveness of vendor-managed inventory in the electronics industry: determinants and outcomes. Information & Management, 41(5): 645–654.
[25] Sari K, (2007) Exploring the benefits of vendor managed inventory. International Journal of Physical Distribution and Logistics Management, 37(7): 529–545.
[26] Vigtil A, (2007) Information exchange in vendor managed inventory. International Journal of Physical Distribution & Logistics Management, 37(2): 131–147.
[27] Boute RN, Disney SM, Lambrecht MR, Van Houdt B, (2007) An Integrated production and inventory model to dampen upstream demand variability in the supply chain. European Journal of Operational Research, 178(1): 121–142.
[28] Chen F, Disney SM, (2007) The myopic order-up-to policy with a feedback controller. International Journal of Production Research, 45(2): 351–368.
[29] Makridakis S, Wheelwright SC, McGee VE, (1998) Forecasting. Methods and applications. West Sussex, UK: John Wiley & Sons,

[30] Dejonckheere J, Disney SM, Lambrecht MR, Towill DR, (2004) The impact of information enrichment on the bullwhip effect in Supply Chains: A control engineering perspective. European Journal of Operational Research, 153(3): 727–750.
[31] Disney SM, Towill DR, (2006) A methodology for benchmarking replenishment-induced bullwhip, Supply Chain Management: An International Journal, 11(2): 160–168.
[32] Deziel DP, Eilon S, (1967) A linear production–inventory control rule'. The Production Engineer, 43: 93–104.
[33] Sterman JD (1989) Modelling managerial behaviour: misperceptions of feedback in a dynamic decision-making experiment. Management Science, 35(3): 321–339.

15

Modelling Diversity Integration in the Organisation

15.1 The Meaning of Diversity in Organisations

In previous chapters related to SC integration we focused on operational and financial aspects of integration. This term, however, can also be extended to specific HHRR programs and activities in the organisation. This is the premise of the current Chapter in which cultural diversity in the organisation will be studied.

Cultural diversity can be defined as the representation in the organisation, as a social system, of people with distinctly different group affiliations of cultural significance [1]. Diversity refers at present to human qualities that are different from our own and those of groups to which we belong, but are manifested in other individuals and groups. Dimensions of diversity include, but are not limited to age, ethnicity, gender, physical abilities/qualities, race, sexual orientation, educational background, geographic location, income, marital or partner status, military experience, parental status, religious beliefs, work experience and job classification.

Nowadays, cultural diversity has become an important issue to pay attention to, and is recently included within the scope of the work of many dynamic modelling analysts in hi-tech companies. The reasoning behind this was, initially, federal regulations requiring plans to increase the participation of minorities and women in the organisation; recently, companies have realised that there is a business case for cultural diversity, and that a diverse and inclusive workplace can become a competitive advantage and increases company profit projections.

15.2 Affirmative Action and Equal Opportunity Policies

In the USA, the term "affirmative action:introduction" (AA) was first introduced by President Kennedy in 1961 as a method of redressing discrimination that had persisted in spite of civil rights laws and constitutional guarantees. It was developed and enforced for the first time by President Johnson. "This is the next and more profound stage of the battle for civil rights," Johnson asserted. "We seek… not just equality as a right and a theory, but equality as a fact and as a result." The term AA describes many policies intended to promote access to education or employment for historically and socio-politically non-dominant groups. In some cases affirmative action involves giving preferential treatment to these underrepresented groups. Nowadays affirmative action is known as just one aspect of the federal government's efforts to ensure equal employment opportunity. An executive order prohibits federal contractors from discriminating against employees on the basis of race, sex, religion, colour, or national origin, and requires contractors to implement affirmative action plans to increase the participation of minorities and women in the workplace. In summary affirmative action:

- targets outreach to underutilised groups;
- helps prevent discrimination;
- is legally mandated;
- measures good faith efforts in making affirmative action progress for minorities and women.

Affirmative action policies also have opponents. They argue that these policies are based on collectivism and that they are merely another equal form of discrimination because they can result in qualified applicants being denied entry to higher education or employment because they belong to a particular social group (usually the historically socio-politically dominant group; typically majority races and men, regardless of social standing or financial need).

Equal Opportunity Employment (often called EOE or EEO) is a term used by the federal government to refer to employment practices that ensure non-discrimination based on race, colour, national origin, sex, physical or mental ability, religion, medical condition, ancestry, marital status or age. The principle behind EEO is that everyone should have the same access to opportunities. This can be achieved because EEO:

- eliminates discrimination in human resource policies and practices;
- provides equal access and opportunity – no one excluded from participation;
- is legally mandated.

Note that AA goes beyond non-discrimination and requires federal contactors to engage in expanded efforts in outreach and recruitment for women and minorities and make them aware of the job opportunities.

The application of these policies and current workforce trends has made diversity become a business issue and a key aspect for gaining competitive advantage.

15.3 A Business Case for Cultural Diversity

Recognising these evolving workplace trends, numerous scholars and business analysts have addressed issues related to diversity in organisations ([1] and [14]).

Their research suggests that firms that effectively manage their workforce diversity may experience positive outcomes. For example, research indicates that:

- firms with higher percentages of women managers report relatively higher financial performance [3] and greater effectiveness [4];
- diversity climate facilitates attracting and retaining (low turnover) a diversity workforce [5];
- diversity atmosphere fosters creativity and innovation in organisations [6–8] and improves organisational flexibility [9,10];
- diversity in marketing groups enhances marketing activities in organisations [9];
- diversity in teams did improve problem solving activities [11–13].

Cultural diversity can thus be seen as a comprehensive organisational and managerial process [2] for developing an environment that maximises the potential of all employees by valuing diversity. Diversity programs in the organisation means therefore one step forward to AA and EEO, involving:

- inclusion of all groups;
- the development of an environment that maximises the potential of all employees by valuing diversity interpersonally and institutionally;
- broader aspects than ethnicity, race, and gender;
- recognition that it is not legally mandated.

Following this path, nowadays in corporations we can find many managers and teams with clear responsibilities in developing diverse actions that may be, somehow, summarised as follows:

- building the business case for diversity in the organisation;
- selling the business case for diversity to current managers, leaders and staff;
- institutionalising diversity through the organisational planning process;
- changing leadership culture to embrace diversity, including training for leadership and key staff ;
- creating a welcoming atmosphere and leadership structure;
- classifying and setting diversity goals;
- measuring and accounting for progress against goals.

15.4 Dynamic Modelling and Cultural Diversity. A Case Study

15.4.1 Purpose of the Modelling Effort

In this case study the organisation was trying to develop a dynamic model with the following purposes:

- identifying the key variables conditioning the reach of the desired diversity in the organisation;
- introducing the diversity learning to date in the model structure;

- evaluating the impact of diversity in business results;
- studying how to proceed for a given business;
- improving learning in diversity.

Note how this is a case where little guidance exists for converting a real-life situation into a simulation model. The team started by trying to understand the problem to model, the model inputs and expected outputs and the boundary definition of the system.

Expected inputs and outputs considered for the model are shown in Figure 15.1. Profit projection of the organisation over time is considered as independent variable for the model and understood as the expected profit in case that the organisational effectiveness would be always 100%.

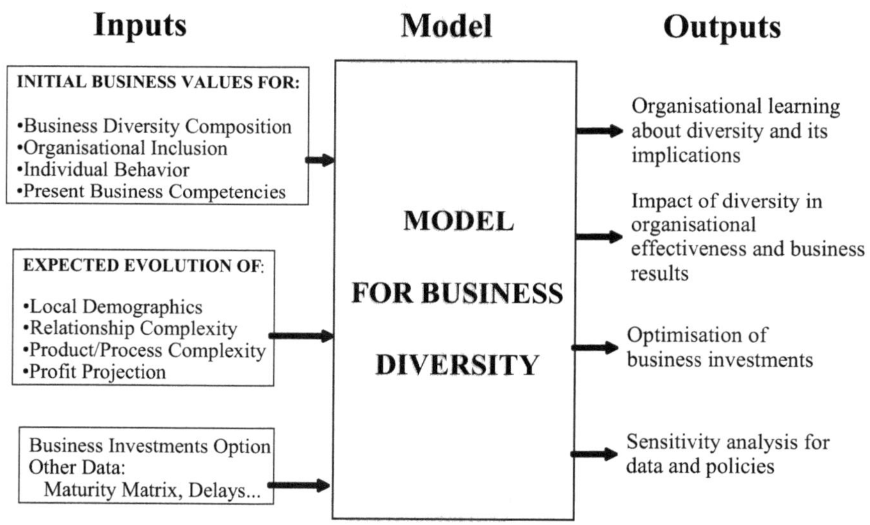

Figure 15.1. Expected inputs and outputs of the diversity model

Another input of the model is a given evolution, according to investment, for the complexity in existing products and processes of the organisation (PPCx) as well as in the complexity of future internal (intra-relationships) and external relationships (inter-relationships) in the organisation (RCx).

The hypothesis introduced in the model is that the increase in PPCx will not be, in relative values, as important as the increase in RCx (see Figure 15.2).

The model considers that demographics are quickly changing towards a higher heterogeneity (see Figure 15.3). The organisations will take this fact into consideration when controlling diversity composition in the future.

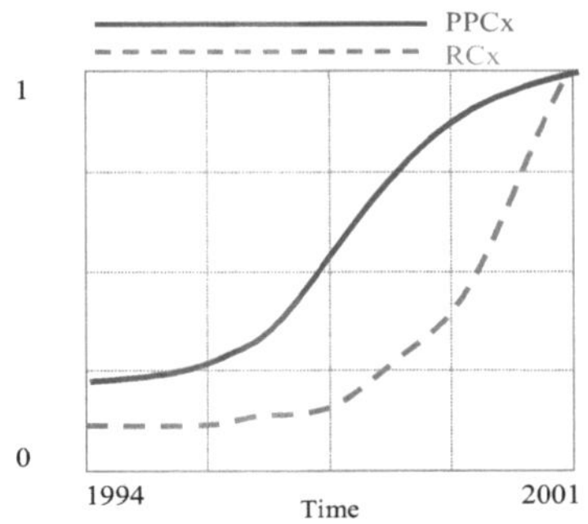

Figure 15.2. Expected evolution of normalized values of PPCx and.RCx

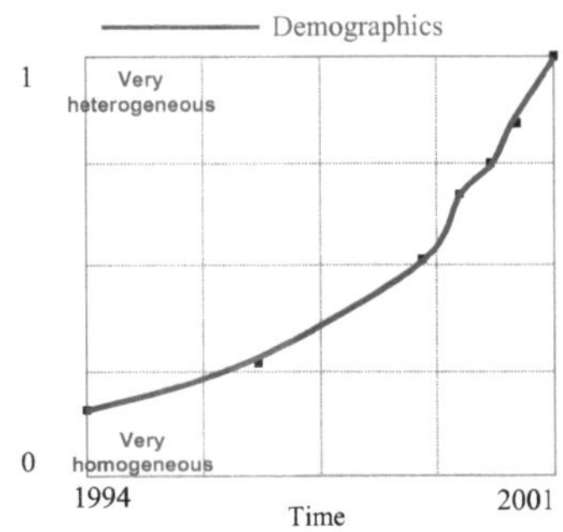

Figure 15.3. Expected evolution of demographics

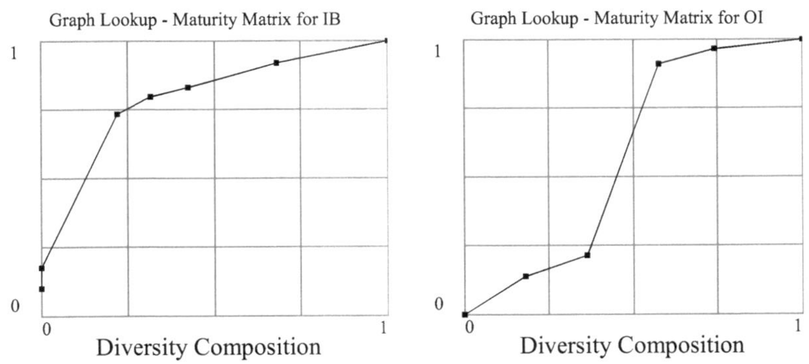

Figure 15.4. IB and OI matrices

As will be explained in the following section, the maturity matrices for individual behaviour (IB) and organisational inclusion (OI) were a very important input for the model, as they capture experience and organisation's estimations (Figure 15.4). These matrices would drive investments and improve diversity competency when required. This information has to be assessed and introduced into the model before running the simulations.

Another very important factor to introduce into the model was the time to recognise the product/process complexity and relationship complexity. There is a basic assumption in the model: a diverse organisation recognises the new complexities earlier, and therefore, it is better prepared to move to a higher level of organisational effectiveness.

15.4.2 Building the Simulation Model

As mentioned before, this is a case where little guidance exists for converting a real-life situation into a simulation model. We did start by identifying the system state variables, or levels in System Dynamics models, and later we developed the flow rates that cause those levels to change, as suggested by Forrester in [15] for these complex modelling situations.

In a first phase, variables determining a key variable for the model, the organisation's *diversity competency,* were considered to be levels of the model:

- diversity composition;
- individual behaviour; and
- organisational inclusion.

For instance, the *diversity composition* (see Figure 15.5) will be trying to reach the desired value set by the organisation, according to local demographics, AA and EEO policies, etc. The increase in composition will be easier with a suitable individual behaviour and a proper organisational inclusion, factors that will, at the same time, produce lower people turnover; diversity composition will not decrease.

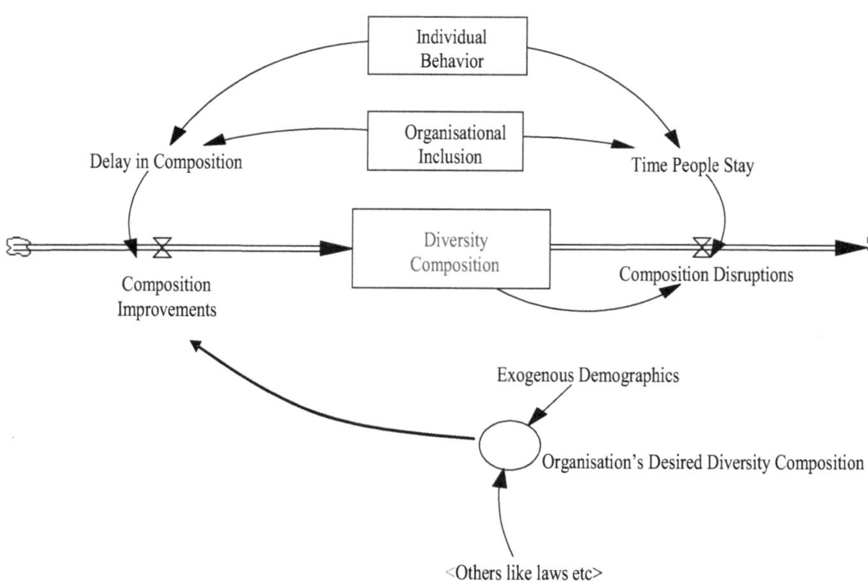

Figure 15.5. Stock and flow diagram of the diversity composition variable

The desired *individual behaviour* (see Figure 15.6) depends on the desired diversity composition, and this relationship was established in the model by the "maturity matrix for IB". This "matrix" was considered a function that, given a desired diversity composition, explains what the minimum required individual behaviour to support is. A diverse composition will increase the exposure to diversity and therefore avoid deterioration of behaviour. Also, organisational inclusion will speed up building behaviour, and reduce this behaviour deterioration.

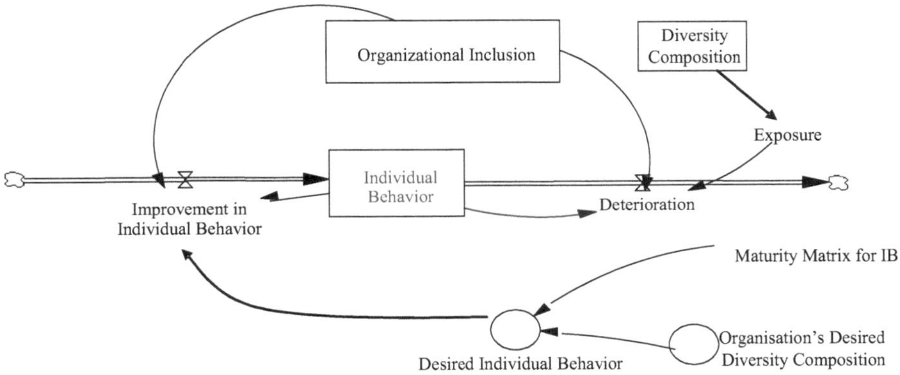

Figure 15.6. Stock and flow diagram of the individual behaviour variable

Desired *organisational inclusion* (see Figure 15.7) was also found to depend on the desired diversity composition. The more diverse the organisation becomes, the more organisational inclusion is required. This relationship was established in the model through the variable "maturity matrix for OI". Again, the level of inclusion reached by the organisation will ease the process of building its own organisational inclusion and avoid its deterioration.

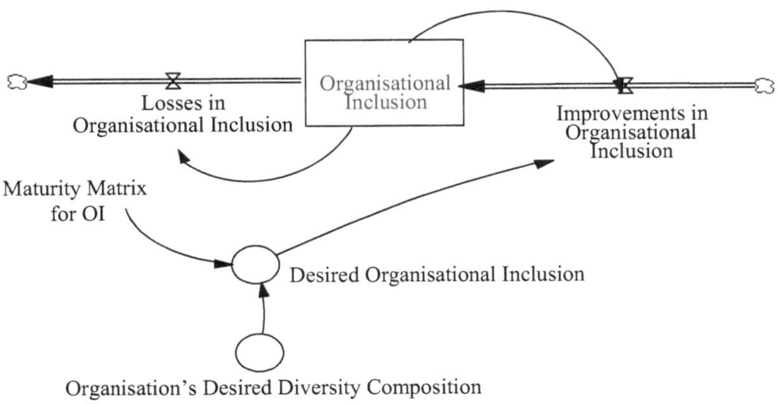

Figure 15.7. Stock and flow diagram of the organisational inclusion variable

In a second phase *diversity competency* was considered by the team as an important factor, a prerequisite to explain and model the future organisation's *competencies in product/process and relationships*. These competencies are assumed to drive organisational effectiveness and therefore were taken as level variables for the model.

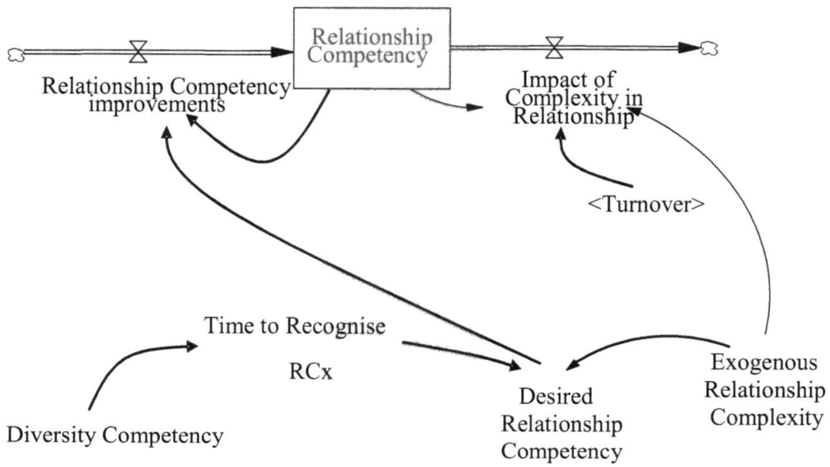

Figure 15.8. Stock and flow diagram of the organisational inclusion variable

Figure 15.8 refers to relationship competency but would also apply to product/process competency. This figure illustrates the fact that when increasing competencies, the organisation looks at the new complexities and establishes the investments required to face them. In the model, diversity competency is assumed to decrease the time required to recognise new complexities. As the reader can see in Figure 15.8, this is required in order to save the impact of the exogenous ("Real" in Figure 15.9) complexity in the competency level and therefore in the organisational effectiveness.

Figure 15.9. Modelling the organisational effectiveness variable

Complexities are exogenous to the model, when they increase, the current competencies of the organisation are not capable of reaching the same effectiveness. In that case, only new investment in competencies may create organisational effectiveness improvements. As we have mentioned, the organisational effectiveness is a result of the organisation's competency in product and process and in relationships. But competency, synonymous of organisation's skills, was finally not considered enough to model organisational effectiveness. At that time the team introduced the variable *individual contribution* in the model, modelled as dependent of current levels of diversity composition, individual behaviour and organisational inclusion (see Figure 15.10).

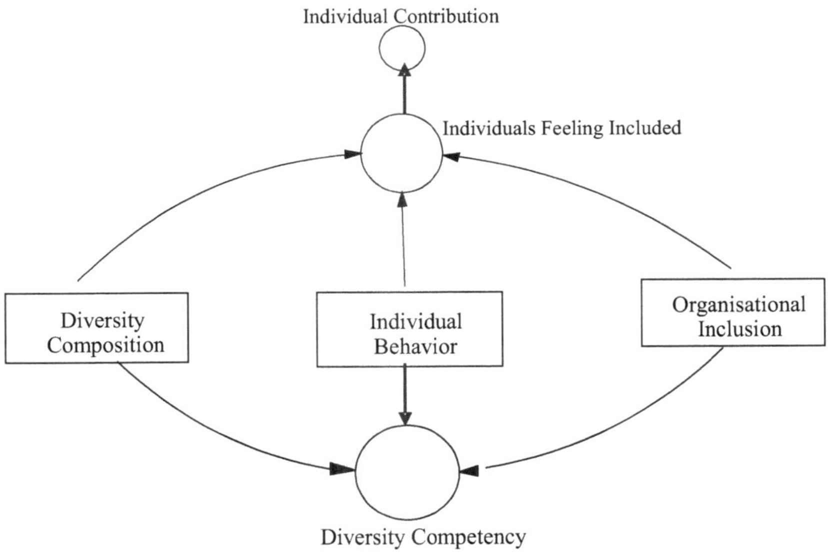

Figure 15.10. Modelling the variable individual contribution

The model obtains the value of business results taking into accounts the profit projection of the organisation, given as exogenous variable, and the calculated organisational effectiveness (see Figure 15.11). The model supposes that a percentage of the business results is dedicated to investments.

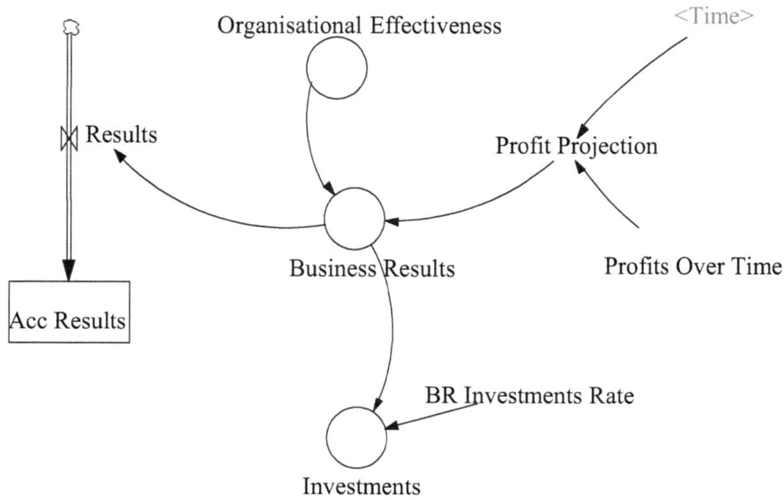

Figure 15.11. Modelling the business results

Causal loops (Figure 15.12) were used for explanation after the model was created and studied, and for brief over-all presentations to people who were not particularly trying to understand the real sources of dynamic behaviour.

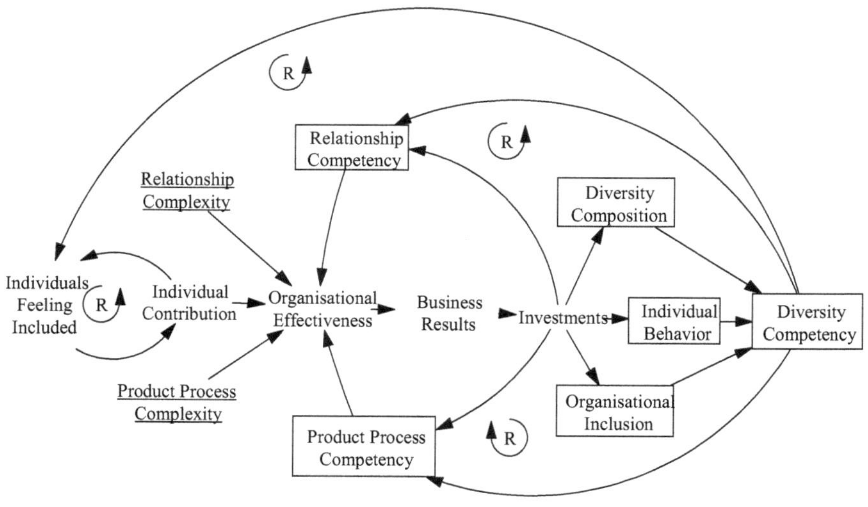

Figure 15.12. Simple overall causal loop diagram

15.4.3 Simulating the Model

Simulation with the model was facilitated for the team members by using a user interface. Thanks to this interface users did not need special skills in System Dynamics to experiment with the model. For instance, there were several menus allowing users to navigate through the model using their mouse and offer them further possibilities for deeper analysis.

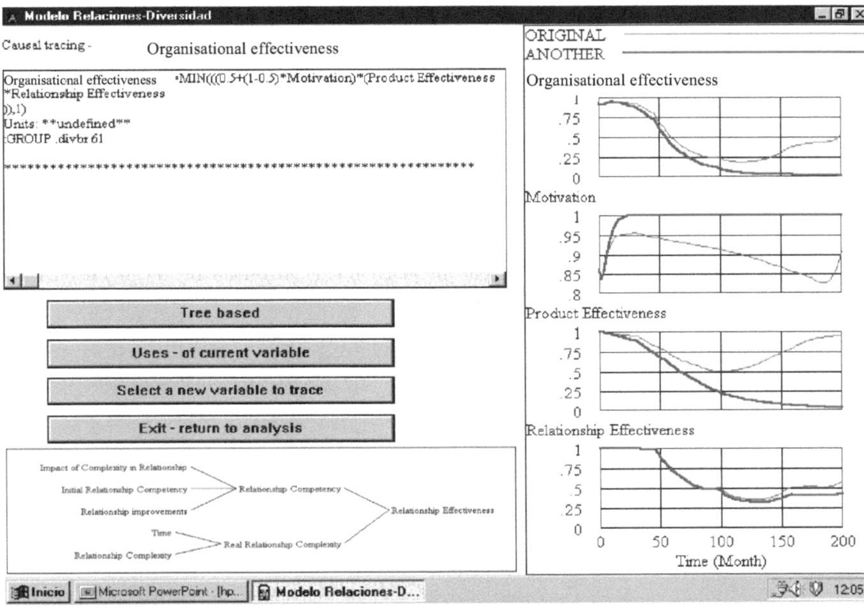

Figure 15.13. Sample model results for two investments options (using the interface)

In the example in Figure 15.13, two runs are compared with a different graphical output, including the equations for the variables. These two runs were produced for the same scenario values, except for the investment policy which is different for each one. The expected results over time are shown.

In Figure 15.14, results for the model variable *organisational effectiveness* are presented. This variable attempts to capture the global impact of diversity on business results. These results were trying to show business managers the importance of considering the different possible investments in diversity in a suitable and appropriate time schedule. A comparison among the results obtained for that variable can be appreciated for the following cases:

- doing nothing;
- investing in product competency only;
- investing in increasing diversity composition only;
- investing in forecasting demographics to improve oi and ib accordingly.

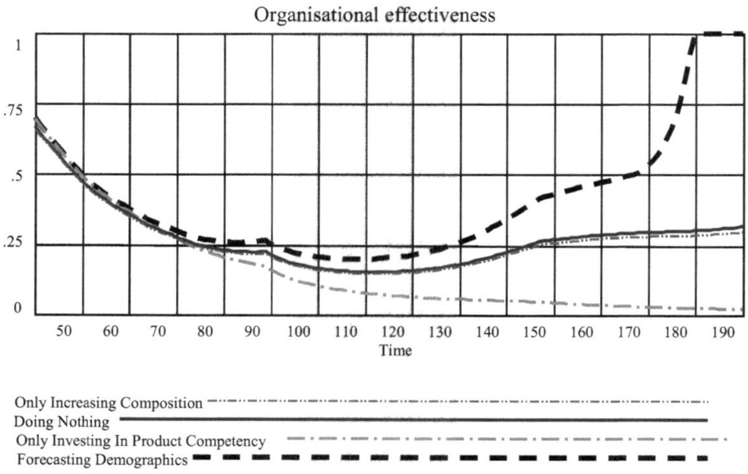

Figure 15.14. Model results for organisational effectiveness and several policy options

Note that in this case study, the model offered the possibility to replicate simulations, changing the parameters values defining the investment options. Optimisation capabilities were embedded in the software package used, and users could replicate simulations automatically until the payoff of the target function did not improve, and its optimum was reached.

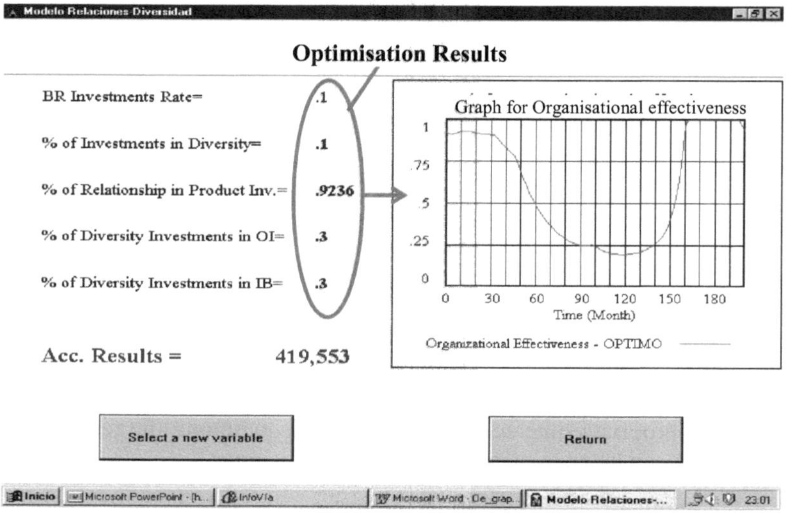

Figure 15.15. Sample model optimisation results

Achieving a global optimum was possible even with high number of parameters to change and for a wide range of parameter variation. The gradient search used by the software was able to accomplish this type of optimisation. In the example in

Figure 15.15 the target function considered was maximising *Accumulated Business Results*. The model offered possibilities to choose the best possible investment policy according to a given scenario definition.

15.4.4 Concluding Remarks of the Case Study

This business case for diversity comprised a dynamic model that was probably the most challenging modelling exercise presented in this book. The project lasted almost 1 year and the people involved in the project team came from different businesses and had very diverse origins and cultural backgrounds. To avoid potential difficulties with the team management, all team members were previously exposed to the *Margerison and McCann* approach [16]. This approach (named *The Team Management Profile*) offers a researched framework of effective team working that helps team members to work both individually and together to realise their full potential. In this special case it was extremely important to learn how the best teams organise themselves to value difference, create balance, enhance critical thinking, set appropriate goals, and improve decision-making.

Regardless of this induction process, team work was not easy. Team members sometimes experienced serious difficulties. For instance, compiling and presenting data became a very difficult task. The results achieved helped the organisation to understand the dimensions of the problem.

The modelling process was recognised to be extremely important. Most of the people participating in the modelling effort agreed that the lessons learnt during the modelling process were even more important for them than the final formal results.

The organisation, by using this model, could plan and schedule suitable diversity investments over time.

15.5 References

[1] Cox TH Jr, (1993) Cultural diversity in organisations: Theory, research and practice. San Francisco, CA: Berrett-Koehler.

[2] Loden M, Rosener JB, (1991) Workforce America! : Managing employee diversity as a vital resource. Homewood, Ill. : Business One Irwin.

[3] Shrader C, Blackburn V, Iles P, (1997) Women in management and firm financial performance: An exploratory study. Journal of Managerial Issues, 9(3): 355–373.

[4] Richard O, Johnson NB, (2001) Understanding the impact of human resource diversity practices on firm performance. Journal of Managerial Issues, 13(2): 177–195.

[5] Konrad A, Linnehan F, (1995) Race and sex differences in line managers' reactions to equal employment opportunity and AA interventions. Group and Organisation Management, 20: 409–439.

[6] Kanter RM, (1977) Men and women of the corporation. New York: Basic Books.

[7] Morgan G, (1989) Endangered species: New ideas. Business Month, 133(4): 75–77.

[8] Nemeth CJ, (1986) Differential contributions of majority and minority influence, Psychological Review, 93(1): 23–32.

[9] Cox TH Jr, Blake S, (1991) Managing cultural diversity: Implications for organisational competitiveness. Academy of Management Executives, 5: 45–56.

[10] Rotter NG, O'Connell AN, (1982) The relationships among sex-role orientation, cognitive complexity, and tolerance for ambiguity. Sex Roles, 8: 1209–1220.

[11] McGrath JE, (1984) Groups: Interaction and performance. Englewood Cliffs, NJ: Prentice-Hall.
[12] Hoffman LR, Maier NRF, (1961) Quality and acceptance of problem solutions by members of homogeneous and heterogeneous groups. Journal of Abnormal and Social Psychology, 62: 401–407.
[13] Shaw ME, (1981) Group dynamics: The psychology of small-group behaviour. New York: McGraw-Hill.
[14] Carter R, (2000) Addressing cultural issues in organisations. Thousand Oaks, CA: Sage.
[15] Forrester JW, (1994) System Dynamics, Systems Thinking, and Soft OR. System Dynamics Review, 10(2–3): 245–256.
[16] http://www.tmsdi.co.uk

Part V

Dynamic Modelling Projects

16

Presenting SCM Dynamic Simulation Projects

16.1 The Project Alternatives

Depending on the supply chain management issue to be dealt with, modellers can be involved in different types of dynamic modelling projects, with different goals, scopes and time horizons. The question that we attempt to answer in this chapter is whether there is any common rule or guideline that we can follow in order to present a dynamic modelling value proposition, especially within the hi-tech industrial sector.

We will now try to offer an orientation for the modeller to follow when approaching a potencial engagement in a modelling project. In order to do so, we will classify dynamic modelling value propositions according to their modelling scope (this classification will be called the *Campbell classification of modelling projects*, in acknowledgement of its original designer, Deb Campbell). Following this criteria, we distinguish four different options or opportunities for the modeller to deliver value to a certain hi-tech organisation. The four options of the Campbell classification are:

- point solution;
- decision improvement process;
- infrastructure solution; and
- organisational independence.

We will now characterise each of these options and then explain how we can develop alternative combinations of them for a specific business.

We have selected the following aspects for a modelling opportunity to be characterised:

- expected result;
- question asked;
- expected deliverables;
- dependency; and
- a sample metaphor.

In the next sections we review each of these aspects for the above modelling engagement options.

16.2 One Point Solution

This type of engagement is characterised as follows:

- *Expected result*: a specific recommendation for a specific question.
- *Question that is asked to the modeller*: how can I fix my specific problem?
- *Expected deliverables*: as a general rule this type of engagement considers two different types of project deliverables:
  - a high-level simulation model based on "what we know today"; and
  - an integrated market document (optional).
- *Dependency*: standalone;
- *A sample metaphor*: the "fish": this work produces a "black box" solution where only the result is used by your customer and his/her organisation.

16.3 Decision Improvement Process

This type of project is characterised as follows:

- *Expected result*: improved organisational understanding and decision making in a target topic area (e.g. the relationship between marketing or manufacturing investments and purchaser behaviour).
- *Question that is asked to the modeller*: how can I improve the effectiveness of my people's decisions when only a few of us understand:
  - the whole business; and/or
  - the impact of independent decisions on our business performance.
- *Expected deliverables*: these deliverables are:
  - a common business context, business language and decision framework;
  - a causal diagram of the target topic area;
  - seminar(s) for the organisation;
  - an integrated market document (optional when standalone, required when a starting phase for "Infrastructure Solution").
- *Dependency*: standalone or point solution;
- *A sample metaphor*: the "gestalt" of fishing. This work does not result in "an answer", but creates a common context, language and decision framework for improved organisational understanding and decision making.

16.4 Infrastructure Solution

This type of project is characterised as follows:

- *Expected result*: recommendations on a variety of questions evaluating the impact of focused investment alternatives, and the tradeoffs between them.
- *Question that is asked to the modeller*: what are the key leverage points that I can use to improve the returns I achieve for the business?
- *Expected deliverables*: these deliverables are:
    - a custom simulation model based on researched data (with varying levels of detail depending on need and availability);
    - an integrated market document (optional).
- *Dependency*: decision Improvement Process.
- *A sample metaphor*: "fishing with assistance": while this work is also focused on providing answers ("fish"), because "Decision Improvement Process" is a pre-requisite, the people involved have a much better understanding of *why* the model results are what they are, and are therefore better prepared to use the results and improve other decisions.

16.5 Organisational Independence

This type of project is characterised as follows:

- *Expected result*: ability for select members of the organisation to run and evaluate model scenarios independently of the modeller;
- *Question that is asked to the modeller*: how can I become independent of the modeller in evaluating investment alternatives?
- *Expected deliverables*: these deliverables are:
    - a user interface;
    - workshop(s) for the organisation.
- *Dependency:* infrastructure Solution;
- *A sample metaphor*: "apprenticeship fishing": this work trains the organisation "to fish" independently. Specific people are trained to run model scenarios, and other key people are educated in how the model can best support their decisions.

16.6 Combination of Alternatives

We can now explain how we can develop alternative combinations of these project options for a certain business. The idea is to show what we can do if the results that we desire go beyond just one of the characterised options.

In Figure 16.1 we can find a visual representation of how these project options can be combined to create a total solution for the business. A potential business interested in modelling projects could enter at any "in" arrow, and exit at any "out" arrow.

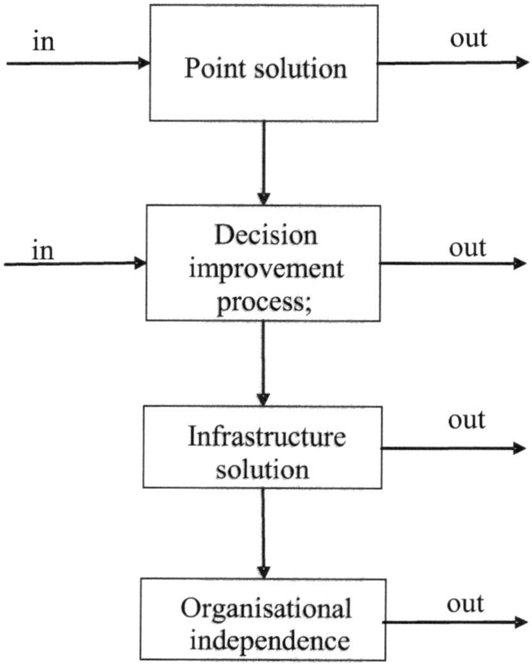

Figure 16.1. A visual representation of the Campbell's classification of projects

Clearly, we cannot start an infrastructure solution or an organisational independence type of modelling project without completing previous phases. The main reason for this is the required organisational understanding at the beginning of the infrastructure solution. A summary of the suggested modelling projects opportunities is offered in Table 16.1.

Table 16.1. Summary of suggested modelling projects opportunities

Problem type	Point solution	Decision improvement process	Infrastructure solution	Org. independence
Desired result	A specific recommendation for a specific question	Improved organizational understanding and decision making in a target topic area (e.g. the relationship between your marketing or manufacturing investments and purchaser behaviour)	Recommendations on a variety of questions evaluating the impact of focused investment alternatives, and the tradeoffs between them	Ability for select members of your organization to run and evaluate model scenarios independently of the modeller
Question asked to the modeller	How can I fix my specific problem?	How can I improve the effectiveness of my people's decisions when only a few of us understand 1) the whole business, and/or 2) the impact of independent decisions on our business performance	What are the key leverage points I can use to improve the returns I achieve for the business?	How can I become independent of the modeller in evaluating investment alternatives?
Deliverables	A high-level simulation model based on what we know today An integrated market document (optional)	A common • Business context • Business language • Decision framework A causal diagram of the target topic area Seminar(s) for the organisation An integrated market document (optional when standalone, required when a starting phase for "Infrastructure Solution")	A custom simulation model based on researched data (with varying levels of detail depending on need and availability) An integrated market document (optional)	A user interface Workshop(s) for the organisation
Dependency	Standalone	Standalone, or point solution	Decision improvement process	Infrastructure solution

16.7 A Modelling Value Proposition. A Case Study

This case study is based on the case presented in Section 5.4 named "Discovery-Driven Planning Process". Based on the issues that were discussed in different meetings with the strategic planning team, we frame together a couple of question to be answered by the dynamic modelling team (see Figure 16.2):

1. Short-term question: "How are we going to make money in this business?";
2. Longer-term question: "How can we increase our confidence in the effectiveness of our decisions?"

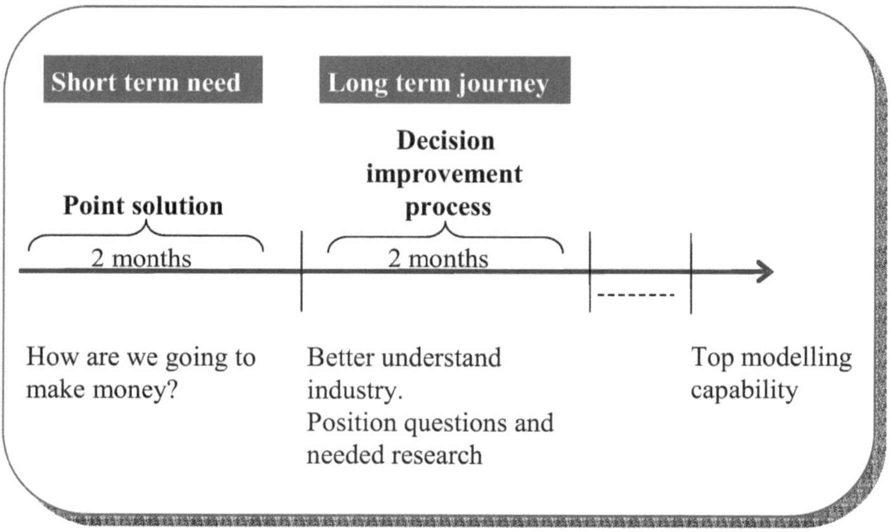

Figure 16.2. Questions asked and timeline for the project

For this case, and according to previous project options, the following alternatives were suggested by the modellers (these alternatives were accompanied by Table 16.2):

Alternative A

- Point solution
- Decision improvement process

Alternative B

- Point solution
- Decision improvement process
- Infrastructure solution

Table 16.2. Elements configuring the alternative options

Result	Deliverables	Time estimate for each result	Funding estimate
Point solution Result: A specific recommendation for the question. What is the most profitable mix of channels for Horizon 1?	A high-level, extensible simulation model based on what we know today and educated assumptions (no detailed research)	2 months	X $
Decision improvement process Result: Improved organisational understanding in Value Delivery Models in the desktop display industry	A common • Business context • Business language • Decision framework A causal diagram of the target topic area Seminar(s) for the organisation An integrated market document (highly recommended for this display business)	2 months	Y $
Infrastructure solution Result: Recommendations on a variety of questions evaluating the impact of focused investment alternatives, and the tradeoffs between them	A custom simulation model based on researched data (with varying levels of detail depending on need and availability) An integrated market document (optional)	2 or 3 months (depends on focus, detail required, and market document)	W $
Organisational independence Result: Ability to select members of your organisation to run and evaluate model scenarios independently of us	A user interface Workshop(s) for the organisation	2 months	Z$

Alternative C

- Point solution
- Decision improvement process
- Infrastructure solution
- Organisational independence

Important assumptions considered in the project, and mentioned in the modelling team business proposition to the business, were:

- Modelling team would focus on channel mix strategy for Horizon 1, (another possibility could be to find out whether to select the System OEM or the branded solution provider. See Figure 16.3 containing a slide of the Point Solution timeline presentation).
- One or two ¼ – ½ time committed people for:
  - Project management;
  - Information gathering and synthesis;
  - Participation in analysis, business implications and recommendations;
  - Capability transfer.
- Access to other people in the organisation for information.
- Team and agreements in place at start date.
- Infrastructure Solution: limited numbers of major research issues arise.

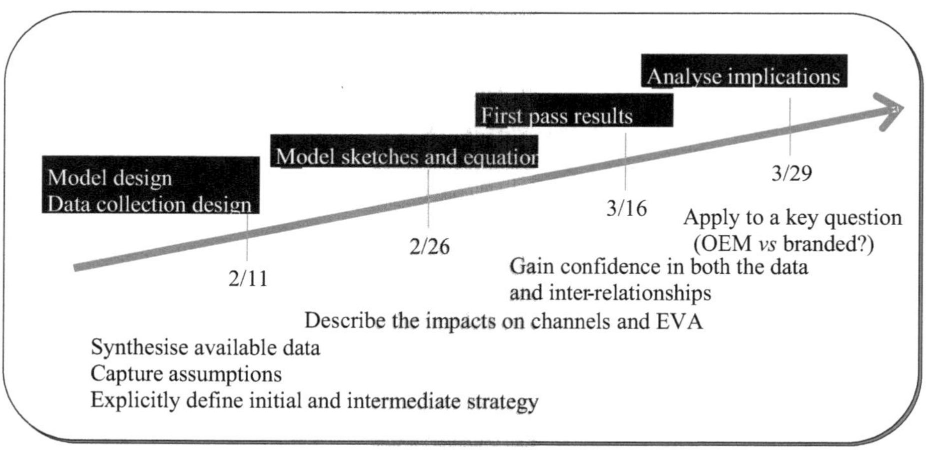

Figure 16.3. Sample Point Solution timeline presentation slide

Other issues mentioned in the business proposition had to do with *cost drivers of the modelling project*. It was mentioned that in this type of modelling projects there are four clear dimensions:

1. project management;
2. technical support;
3. data collection and synthesis; and

4. Application of the results into the organisation.

The costs for the modelling work was driven by:

- the complexity of these dimensions;
- how much of each dimension the modelling team needs to lead or do; and
- how fast the results are expected.

In summary, the major cost drivers for this project were:

- desired speed of result;
- assumptions about data or organisational complexity;
- assumptions about people availability from the business;
- assumptions about information availability from the business.

17

Capturing the Learning of a Modelling Project

17.1 The Project Technical Closure

Best practices in project management (see PMBoK [1] [1]) take into acount the importance of the technical, besides the administrative, closure of the project. Future similar projects will benefit from the time and effort that the modeller put into the modelling work if he/she is able to assimilate properly the lessons learned during the entire life cycle of the project. Therefore, this task should be a must for a world class modeller. Closure of the project will normally require performance reporting. This reporting activity will take into consideration certain inputs and will use tools and techniques to produce specific outputs. This closure procedure can be summarised as follows:

- Inputs required to do this performance reporting in projects and in general are [1]:
  - the project plan;
  - the work results;
  - other project records.
- Tools and techniques that can be used are [1]:
  - performance reviews;
  - variance analysis;
  - trend anlaysis;
  - earned value analysis;
  - information distribution tools and techniques.

---

[1] "PMBoK" is a trademark of the Project Management Institute, Inc. Which is registered in the United Status and other nations.

Table 17.1. Suggested Technical Completion Document

Model purpose and strategy
- Modelling approach used: What alternatives were considered? How did this approach more appropriately address the core issues? - How was the model used to address the project objectives? What analysis was complete? - How was the analysis used, or if not used, how did the team recommend it be used? - What important limitations were encountered in developing or using the model? - Are there additional uses of this model? Can it be generalised?
Archives files and documents
- Describe the files used in the modelling work and the convention to identify file versions. - How is each file used? - What files must be updated and what agreements were made?
Model structure
- Describe the general structure of this model. *Provide a brief description of each variable (significance to problem, important relationships to other variables, choice of level of detail)* *Describe or define any subscripts that were used. How did this added detail add to the model's accuracy or quality?* - Describe the important assumptions used in building this model? - Describe the approach used to validate the model. *How would you defend the assumptions and the use of this model to a technical audience (standard validation approaches, fit with historical data)?* *How would you do the same for non-technical decision-makers?*
Model use
- Describe how to set up and use this model (assume the user is familiar with the software, data and data files). *What data must be updated and loaded?* *What scenarios or optimisations must be completed? How is this done?* *How does the user set up and/or access output files?* *How should the output be interpreted?* - Using information from model archives, what useful examples exist to illustrate conclusions that were drawn from the output of prior model runs and analysis?
Maintenance
- Do any data files require updating? - If so, what can be or should be gleaned from future model runs using updated data? - How can or should the model be updated to improve its usefulness?
Technical learning
- What important technical lessons came from this modelling effort (new approaches, improvements to existing approaches)? - What significant technical problems were identified and solved to complete this analysis? - What benefits (if any) did the technical team get from their involvement in the project?

- Outputs that can be produced are [1]:
  - performance report;
  - change requests;
  - project archives;
  - lessons learned.

We can adapt this type of output to our dynamic modelling project and design a specific technical completion document (TCD) for it. In this book we are proposing a structure for the TCD as presented in Table 17.1.

17.2 The Project Technical Closure Case Study

In this case study the TCD structure presented above is applied to a dynamic simulation model that is presented in Chapter 7. The referred model tried to formalise how customers behave with regard to new products and services that came out in different periods of time. By understanding this point the project team could help the business by figuring out a strategy for the product to be able to meet their needs and demand and also cater to their preferences.

The following TCD for this case study attempts to capture the learning achieved when modelling how the customer perception of product quality and price attributes was impacting market share for a certain company and for its competitors.

17.2.1 Model Purpose and Strategy

- *Describe the modelling approach used. What alternatives were considered? How did this approach more appropriately address the core issues?*

  The approach used is System Dynamics. SD facilitates the formalisation and visualisation of the customers behaviour by structuring the decision making process and by linking that, afterwards, to the financial variables. Understanding customer behaviour, the marketing investment alternatives can be leveraged using financial suitable metrics.

- *How was the model used to address the project objectives? What analysis was complete? How was the analysis used, or if not used, how did the team recommend it be used?*

  The model was providing the best alternatives for attributes investments according to the segment teams requisitions. Also the model was offering the global effect of those alternatives — for all the segments at the same time — and the possible best combination of attributes to invest in.

  The modelling team recommended using the model once the possible investments in product attributes had been deeply explored: what changes in attributes perceptions were really feasible and while were not. Otherwise the process turns into one of "trial and error".

- *What important limitations were encountered in developing or using the model?*

At this stage of development, the model does not take into account the cost of an investment for an attribute change, but provides its impact on revenue. The user has to wonder then whether the proposed investments are worthwhile or not.

- *Are there additional uses for this model? Can it be generalised?*

  The model could be used to explore the reseller attributes too — this was done in several cases . Moreover, the way the perceived customer value is modelled can be generalised for many other modelling purposes, for example, modelling the probability of a supplier to allocate capacity.

17.2.2 Archives, Files and Documents

- *Describe each of the files used in the modelling work. What convention was used to identify file versions or dates?*

  There is only one model file, but there are several files if the user interface is used. For this particular project the interface was used only for presentation purposes, since a specific partner in the business has ample knowledge of the modelling software application and its capabilities.

- *How is each file used? What files must be updated, and what agreements were made regarding maintenance and updates?*

  The only file to be updated is the spreadsheet which captures the market research data that is introduced into the model, automatically when the model imports the data file.

17.2.3 Model Structure

- *Describe the general structure of this model.*

  a) Provide a brief description of each variable (significance to problem, important relationships to other variables, choice of level of detail).

  Main variables are: customer perception of product attributes relative to competition and price of the product. These are two subscripted level variables conditioning the rest of the model variables – for example, financial variables.

  Main parameters are importance of each attribute for the customer — which is used to obtain the perceived value for each segment, table for value and probability — which based in historical data links value to probability of purchase, usage rate — which changes differently for each segment over time, segment size — also changes over time, and currency exchange (Yen) — historical data and forecast.

  b) Describe or define any subscripts that were used. How did this added detail add to the model's accuracy or quality?

Segments of customers, segments of printers, site sizes, quality attributes and within them quality sub-attributes and price attributes;

In this particular model, the price analysis had to be done per product segment, while revenue main analysis was done per customer segment. Sub-attribute level for quality factors is a main feature of the model which helps one to be more precise in the study of possible investments.

- *Describe the important assumptions used in building this model.*

The main assumption can be considered to be the fact that value perception is driving the probability of purchase assuming the same correlation over the years. Also in analytical terms, the way that we obtain the relative values to the competition could be an aspect to improve, but the method used in this model is a general market practice.

- *Describe the approach used to validate the model.*

First of all, the structure was initially validated by the team members. Second, the perceived value was obtained for each segment and compared with the market share information from the research data, and for the different site size. Also the prices for product segment and the rest of the financial data was validated.

a) How would you defend the assumptions and the use of this model to a technical audience (standard validation approaches, fit with historical data)?

The main assumption seems to be reasonable for the time horizon of the model. The rest of the assumptions are data provided by the customer.

b) How would you defend the assumptions and use of this model to non-technical decision-makers?

By showing the structure of the model as well as the the output data that the model produces and comparing them to the real data.

17.2.4 Model Use

- *Describe how to set up and use this model (assume the user is familiar with the software, data and data files).*

The user has to set up a strategy to change the attributes perception over the years and characterise the customers segments according to the importance they pay to the different attributes. Then the segments size and usage rates have to be entered. That data will be enough for the model to produce market share output to validate. Once we have the share validated we can check that the prices are behaving as expected, translating correctly share in customer segments into share in product segments. Finally the user can obtain and compare the financial results.

Once the user follows this process the first time, he/she will only need to change strategies for investments later. And analyse results conveniently.

a) What data must be updated and loaded?

This is described in the previous paragraph.

b) What scenarios or optimisations must be completed? With some level of specificity, how is this done?

Regarding optimisation, the optimisation capabilities of the software used have been considered in order to select the more suitable attributes to invest in, according to a payoff function of the user choice. Different examples have been created to optimise, for instance, the share in a segment, the unit sales, or the total revenue.

c) How does the user set up and/or access output files?

Output reports and Vensim data files are available at any time.

- *Using information from model archives, what useful examples exist to illustrate conclusions that were drawn from the output of prior model runs and analysis?*

  Check project slides.

17.2.5 Maintenance

- *Do any data files require updating?*

  All data files have to be updated according to new market research data and financial information.

- *If so, what can be or should be gleaned from future model runs using updated data? How can or should the model be updated to improve its usefulness?*

  A practical and simple issue is to connect all the data in the spreadsheet automatically into the model using data variables. Another possibility is to connect the market sharing functions used in this model to other business activities like, for example, understanding suppliers loyalty

17.2.6 Technical Learning

- *What important technical learning came from this modelling effort (new approaches, improvements to existing approaches)?*

  The fact is that the marketing estimations can be produced with more robustness and rigor based on the marketing functions that we use to define the value proposition for the customer.

- *What significant technical problems were identified and solved to complete this analysis?*

  Make sure that the marketing function used to estimate the customer value perception sorts the segments in the same order that the market research does. Also connect customer segment market share with product segment market share which drives the price of the product's consumible.

- *What benefits (if any) did the technical team get from their involvement in the project?*

  The principal benefit was seeing how we have built the capability to answer customer questions in a much more structured way, avoiding typical "guesstimations" in this field. Another benefit is the fact that the model ownership is ensured in the organisation. The model is "alive".

17.3 Reference

[1] PMBoK, PMI Standards Committee, (1996) A guide to the project management body of knowledge. Four Campus Boulevard. Newtown Square PA. USA

Index

MIX
Papier aus verantwortungsvollen Quellen
Paper from responsible sources
FSC® C105338

If you have any concerns about our products,
you can contact us on
ProductSafety@springernature.com

In case Publisher is established outside the EU,
the EU authorized representative is:
Springer Nature Customer Service Center GmbH
Europaplatz 3, 69115 Heidelberg, Germany

Printed by Libri Plureos GmbH
in Hamburg, Germany